EXPOSURE

EXPOSURE

Poisoned Water, Corporate Greed and One Lawyer's Twenty-Year Battle against DuPont

ROBERT BILOTT

with Tom Shroder

**SIMON &
SCHUSTER**

London · New York · Sydney · Toronto · New Delhi

A CBS COMPANY

First published in the United States by Atria Books, an imprint of
Simon & Schuster, Inc., 2019
First published in Great Britain by Simon & Schuster UK Ltd, 2019
A CBS COMPANY

1 3 5 7 9 10 8 6 4 2

Simon & Schuster UK Ltd
1st Floor
222 Gray's Inn Road
London WC1X 8HB

www.simonandschuster.co.uk
www.simonandschuster.com.au
www.simonandschuster.co.in

Simon & Schuster Australia, Sydney
Simon & Schuster India, New Delhi

A CIP catalogue record for this book is available from the British Library.

Hardback ISBN: 978-1-4711-8961-6
Trade Paperback ISBN: 978-1-4711-8962-3
eBook ISBN: 978-1-4711-8963-0

Interior design by Dana Sloan

Printed in the UK by CPI Group (UK) Ltd, Croydon, CR0 4YY

MIX
Paper from
responsible sources
FSC
www.fsc.org   FSC® C020471

For my sons, Teddy, Charlie, and Tony

They wanna try and keep everything hushed up. Like it's some kind of big secret. . . . They won't tell us what it is. They don't wanna talk to me. Because I'm an old dumb farmer, I'm not supposed to know anything. But it's not gonna be covered up. Because I'm gonna bring it out in the open for people to see.

—Wilbur Earl Tennant

Author's Note

Facts and events set forth in this book are based on and a continuation of my efforts over the last several decades to warn of the past and continuing dangers to the public arising from the behavior of E. I. du Pont de Nemours and Company and others presenting a likelihood of substantial harm to individuals and the public interest arising from exposure to PFOA and related PFAS chemicals. Such facts and events are based on and/or derived from information, documents, and/or testimony and/or statements available in public court filings, trial transcripts, regulatory agency or other public dockets or document repositories and/or publications, media statements, and/or other means by which information is publicly available or otherwise contained in or available through a public record and/or not subject to any court-approved confidentiality or protective-order restrictions. All names and characters are unchanged. Out of an abundance of caution, publication of this book comes only after several public jury trials, including three multiple-week-long jury trials in federal court in Ohio resulting in final jury verdicts and orders, including punitive damage verdicts, and completion of full appellate briefing and oral argument. Certain dialogue and events are reconstructed to the best of the author's ability, with every effort undertaken to avoid revealing or disclosing any attorney-client advice, confidences, or work product and without waiving in any way any such privilege, protection, or other legally recognized or otherwise available privileges or protections for any client or other person, and no person shall interpret anything in this book as representing or constituting any such waiver, as any and all such alleged waivers are expressly denied. All opinions and dialogue of the author (outside of quoted court transcripts) are his purely personal opinions, recollections (to the best of his ability), feelings, and statements, and are not to be attributed to any current or former client, partner, law firm, or other person for any purpose.

Contents

Act I

THE FARMER

1

DRY RUN

July 7, 1996
Washington, West Virginia

No one would help him.

The cattle farmer stood at the edge of a creek that cut through a sun-dappled hollow. Behind him, white-faced Herefords grazed in rolling meadows. His mother's grandfather had bought this land, and it was the only home he had ever known. As a boy, he had cooled his bare feet in this creek. As a man, he had walked its banks with his wife. As a father, he had watched his little girls splash around in its shallow ripples. His cattle now drank from its pools.

The stream looked like many other streams that flowed through his sprawling farm. It was small and ephemeral, fed by the rains that gathered in the creases of the ancient mountains that rumpled West Virginia and gave it those misty blue, almost-heaven vistas. Thunderstorms occasionally swelled the creek so much that he couldn't wade across it. Dry spells shrank it to a necklace of pools that winked with silver minnows. Sometimes it ran so dry he'd find them glittering dead in the mud. That's why they called it Dry Run.

Dry Run used to flow gin clear. Now it looked like dirty dishwater. Bubbles formed as it tumbled over stones in a sudsy film. A thicker foam gathered in eddies, trembling like egg whites whipped into stiff peaks so high they sometimes blew off on a breeze. You could poke it with a stick and leave a hole. It smelled rotten.

"That's the water right there, underneath that foam," the farmer said.

He was speaking to the camcorder pressed to his eye. No one believed him when he told them about the things he saw happening to his land. Maybe if he filmed it, they could see for themselves and realize he was not just some crazy old farmer. Birds sang through the white-hot humidity as he panned the camcorder across the creek. His hand shook as he pressed the zoom button, zeroing in on a stagnant pool. Its surface was matte with a crusty film that wrinkled against the shore.

"How would you like for your livestock to have to drink something like that?" he asked his imagined audience.

The farmer's name was Wilbur Earl Tennant. People who didn't know him very well called him Wilbur, but friends and family called him Earl. At fifty-four, Earl was an imposing figure, six feet tall, lean and ox-shouldered, with sandpaper hands and a permanent squint. He often walked through the woods shirtless and shoeless, his trousers rolled up, and he moved with an agile strength built by a lifetime of doing things like lifting calves over fences.

Hard labor was his birthright. It had paid for the 150 acres of land his great-grandfather had bought and for the two-story, four-room farmhouse pieced together from trees felled in the woods, dragged across fields, and raised by hand. The farmhouse stood at the foot of a sloping meadow that rose into a bald knob.

Dry Run was less than a mile's walk from the home place, across Lee Creek, through an open field, and along a pair of tire tracks. It flowed through a corner of the three-hundred-acre farm, in a place Earl called "the holler." A small valley cut between hillsides, the holler was where he moved the herd to graze throughout the summer. He walked there every day to count heads and check fences. The cows grazed on a mixed pasture of white Dutch clover, bluegrass, fescue, red clover . . . "just a duke's mix of everything." Until lately, the cattle always fattened up nicely on that, plus the corn he grew to finish them and a grain mix he bought from the feed store. Now, he was feeding them twice as much and watching them waste away.

The problem, he thought, was not what they were eating but what they were drinking. Sometimes the cattle watered at a spring-fed bathtub trough at the farthest end of the field, but mostly they drank from Dry Run. Earl had come to believe that its water was now poisoned—with what, he did not know.

"That's where they're supposed to come down here and pull water samples, to see what's in that water." He pointed the camera at a stagnant pool of water flanked by knee-high grass. The olive green water had a greenish brown foam encrusting the grassy bank. "Isn't that lovely?"

The edge in his voice was anger. His cattle were dying inexplicably, and in droves. In less than two years he had lost at least one hundred calves and more than fifty cows. He marked each one on a calendar, a simple slash mark for each grotesque death. The herd that had once been nearly three hundred head had dwindled to just about half that.

Earl had sought help, but no one would step up. After contacting the West Virginia Division of Natural Resources and the West Virginia Department of Environmental Protection, he felt stonewalled. He said the state vet wouldn't even come out to the farm. He knew the folks at the DNR, because they gave him a special permit to hunt on his land out of season. But now it seemed they were ignoring him.

"It don't do you any good to go to the DNR about it. They just turn their back and walk on," he told the camera. "But you just give me time. I'll do something about it."

Thing was, time was running out. It wasn't just his cattle dying. Deer, birds, fish and other wildlife were turning up dead in and around Dry Run. He had stopped feeding his family venison from the deer he shot on his land. Their innards smelled funny and were sometimes riddled with what looked to him like tumors. The carcasses lay where they fell. Not even buzzards and scavengers would eat them.

Hunting had been one of Earl's greatest pleasures. He had carried a rifle as he went about the farm, always ready to shoot dinner. He was an excellent marksman, and his family had always had enough meat to eat. His freezer had brimmed with venison, wild turkey, squirrel, and rabbit.

Now it was filled with specimens you might find in a pathology lab.

The problem had to be Dry Run, he thought. He hardly ever saw minnows swimming in the creek anymore, except the ones that floated belly up.

He zoomed out and panned over to an industrial pipe spewing froth into the creek.

"But the point I want to make, and make it real clear," he said, zooming in, "that's the mouth of Dry Run."

The pipe flowed out of a collection pond at the low end of a landfill. On the other side of his property line, Dry Run Landfill was filling up the little valley that had once belonged to his family. Its dumping pits were unlined, designed for the disposal of nonhazardous waste—office paper and everyday trash. At least that's what his family had been told thirteen years before by the company that had bought their land.

He didn't believe it anymore. Anyone could see that something was terribly wrong, not only with the landfill itself but with the agencies responsible for monitoring it. Did they think no one would notice? Did they think he would just sit by?

"Somebody's not doing their duty," he said to the camera, to anyone who would listen. "And they're going to find out one of these days that somebody's tired of it."

. . .

Two weeks after he filmed the foamy water, Earl aimed the camcorder at one of his cows. Standing walleyed in an open field was a polled Hereford—red with a white face and floppy ears.

He panned the camera a few degrees. Her calf, black and white, lay dead on its side in a circle of matted grass. It looked, at most, a few days old. Its head was tipped back at an awkward angle. The carcass was starting to smell.

"It's just like that other calf up yonder," he said, panning over the matted grass. "See how that's all wallered down? That calf had died miserable. It kicked and thumped and wallered around there like you wouldn't believe."

The calf was engulfed in a black, humming mist. He zoomed in. Flies.

He panned again: a bonfire on a grassy slope, a pyre of logs as fat as garbage cans. In the flames, a calf lay broadside, burning. Black smoke curled into the daylight.

"This is the hundred and seventh calf that's met this problem right here. And I burn them all. There's been fifty-six cows that's been burnt just like this."

In another field, a grown cow lay dead. Her white hide was crusted with diarrhea, and her hip bones tented her hide. Her eyes were sunk deep in her head.

"This cow died about twenty, thirty minutes ago," Earl said. "I fed her at least a gallon of grain a day. She had a calf over there. Calf born dead."

Earl loved his cows, and the cows loved Earl. They would nuzzle him as he scratched their heads. In the spring, he would run and catch the calves so his daughters could pet them. Even though he sold them to be finished and slaughtered for beef, he didn't have the heart to kill one himself, unless it had a broken leg and he needed to end its suffering. Recently, the cows had started charging, trying to kick him and butt him with their heads, as this one had before she died.

She had spent the summer in the hollow, drinking out of Dry Run until she'd started to act strangely. With no one from the government or even local veterinarians willing to do it, Earl decided to do an autopsy himself. It wasn't his first. His earlier efforts had all revealed unpleasant surprises: tumors, abnormal organs, unnatural smells. He wasn't an expert, but the disease seemed clear enough that he bagged the physical evidence and left it in his freezer for the day he could get someone with credentials interested enough to take a look. That day had never come, so he decided he would make them watch a video.

"She's poor as a whippoorwill. And I'm gonna cut her open and find out what caused her to die. Because I was feeding her enough feed that she shoulda gained weight instead of losing weight. The first thing I'm gonna do is cut this head open, check these teeth."

Earl pulled on white gloves and pried open the cow's mouth, probing her gums and teeth. The tongue looked normal, but some of the teeth were coal black, interspersed with the white ones like piano keys. One tooth had an abscess so large he reckoned he could stick an ice pick clear under it. The flies hummed as loud as bees. He sliced open the chest cavity, pulled out a lung, and turned the camera back on. The smell was odd. He couldn't quite place it. It was different from the regular dead-cow smells he had dealt with all his life.

"I don't understand them great big dark red places across there. Don't understand that at all. I don't ever remember seeing that in there before."

He cut out the heart and sliced it open. The muscle looked fine, but a thin, yellow liquid gathered in the cavity where it once beat. "There is about a teacup or so full of it—it's a real dark yeller. It's something I have never run into before."

He reached back into the cow and pulled out a liver that looked about right. Attached to it was a gallbladder that didn't. "That's the largest gall I ever saw in my life! Something is the matter right there. That thing's about . . . oh, two-thirds bigger than it should be."

The kidneys, too, looked abnormal. Where they should have been smooth, they looked ropy, covered with ridges. The spleen was thinner and whiter than any spleen he had come cross. When he cut out the other lung, he noted dark purple splotches where they should have been fluffy and pink. "You notice them dark place there, all down through? Even down near the tips of it. That's very unusual. That looks a little bit like cancer to me."

Whatever had killed this cow appeared to Earl to have eaten her from the inside out.

2

THE CALL

October 9, 1998
Cincinnati, Ohio

My office phone rang.

"Hello," said an unfamiliar voice. "Is this Robbie?"

"This is Rob Bilott," I said, startled. No one had called me Robbie in twenty years.

The man spoke with a distinct southern drawl, the nasally twang of Appalachia, and his words ran together in one breathless run-on sentence. He was highly agitated, talking so fast that I struggled to understand him. The man needed help. Something about . . . cows? Not just cows but dying cows. I was a corporate defense lawyer. Why was he calling me? The volume and tone of his voice set me back in my chair. How did this guy get my number? I rarely got calls on my private line from anyone besides clients, other lawyers, and family. Anyone else would have called our main office; the number for my direct line was unpublished. As he rattled on about his cows, I debated transferring him to my secretary or simply hanging up. Then he caught my attention with two words: "Alma White."

Grammer? What could this dying-cow story possibly have to do with my eighty-one-year-old grandmother?

9

I reached for a legal pad and wrote down his name: Earl Tennant. His family owned a cattle farm outside Parkersburg, West Virginia—the town where Grammer had lived for years. This was my mom's hometown, too. She had grown up in Parkersburg and had attended Parkersburg High School. Whenever we packed up the car to head to West Virginia when I was a kid, she always referred to it as "going home."

It was a town I knew well from those childhood trips to visit Grammer and my great-grandparents. I had even lived there briefly, in elementary school. I was a military brat, moving from base to base, leaving new friends as soon as I'd made them. Parkersburg was one of the only constants of my young life, the source of many of my happiest childhood memories. When I think of an idyllic Christmas, Parkersburg is where it's set. I had a small family, with just my parents and one sibling: my one-year-older sister, Beth. But we'd have huge dinners with my grandmother and great-grandparents, then drive out to tour the modest fifties and sixties ranch houses in Grammer's neighborhood, made glorious with over-the-top Christmas lights and decorations.

Though close by, Earl's farm was a world away from Grammer's semisuburban neighborhood, somewhere in the hills south of town, next to a farm run by a family by the name of Graham. Earl had been telling his neighbor, Ann Graham, about his dying cows. He told her that chemicals were leaking from a nearby landfill and poisoning his herd. He'd tried local attorneys, but they had recoiled once they'd heard who owned the landfill—DuPont, the biggest employer in town—and curtly told him to find another lawyer. The "government folks" weren't any help, either. They listened to him rattle on, then did nothing. He'd called DuPont more times than he could count. He said the people who'd answered the phone had given him slippery responses. This had been going on for two years as his cows kept dying. He told his neighbor he was going to find a lawyer, even if he had to hunt one down in another state.

Ann Graham wanted to help. It happened that she'd just been on the phone with my grandmother, who now lived near my parents in Dayton, Ohio. Grammer had been bragging about my working as an environmental lawyer for a fancy-pants firm in Cincinnati. Through Ann,

Grammer gave Earl my name and number along with a glowing recommendation born more of grandmotherly pride than any real legal understanding. She had promised that surely her grandson would help.

I silently shook my head. Grammer should have asked me before handing out my number like a Halloween treat. I would have patiently explained that I was exactly the wrong kind of environmental lawyer—my firm did corporate defense for folks just like DuPont when they were being accused of polluting someone's farm. In short, I was playing for the opposite team.

But I decided to hear the man out. The pain and passion in his voice moved me. I tried to imagine standing by helplessly and watching your animals suffer and die while nobody lifted a finger to help. The last thing I wanted to do was add to the litany of people and institutions that had failed him. I wanted to help him. But I had questions.

"Do you have documents that prove DuPont owns the landfill?"

"Yep," he said. In fact, he and his family had sold DuPont some of the land that was now Dry Run Landfill. He still had the papers.

"How do you know that chemicals from the landfill are getting into your cows?"

"I've dissected several of 'em. I've never been into anything like it in my lifetime," he said.

"Do you have any idea what exactly is leaking out of the landfill?"

"No," he said, but he was absolutely certain there was something. Something bad. It had to be to make his herd this sick. What's more, he believed, someone knew it and was trying to keep it quiet. Why else would everyone just keep looking the other way?

I was skeptical. You might say I'm professionally skeptical. DuPont was not some fly-by-night company operating under the radar of the regulators. They were one of the biggest chemical corporations in the world, nearly two hundred years old and highly respected, a company widely considered an industry leader in health and safety at the time.

I knew this because, though they weren't a client, I had worked side by side with their in-house environmental counsel at various Superfund sites—the government-mandated effort to clean up some of the country's most contaminated land and water. They were top-shelf lawyers. They had the training—and the budget—to do things right, which in my experience was what corporate environmental lawyers always wanted to do.

Earl's insistence that DuPont was secretly dumping hazardous chemicals into a landfill that was permitted only for nonhazardous waste sounded a little too much like a conspiracy theory for my taste. In my years dealing with my firm's corporate clients, I had never encountered that kind of behavior. Earl must have heard the doubt in my voice.

"I can prove all this," he insisted. He had boxes of photos and videotapes documenting everything he was describing. If I could just see them, I'd understand, he told me. "They'll make it all crystal clear."

What was already clear was this: if what Earl said was true, it would mean a possible lawsuit against one of the world's biggest chemical firms.

I didn't have the heart to tell Earl that I didn't sue corporations like DuPont, I defended them. Their fees paid my bills.

But I've always hated letting anyone down, especially someone in a desperate situation. I figured the least I could do was invite him to the firm, hear him out, and look at whatever he so desperately wanted to show me. Besides, he was calling at the suggestion of my grandmother. I invited him to make the three-and-a-half-hour drive to Cincinnati. He could bring all his boxes. If nothing else, I could refer him to an attorney at another firm who did this kind of work.

"I can't make any promises," I said. "But I'll take a look."

Worst case, I'd waste an hour looking at a random collection of photos and documents that didn't add up to much.

How much trouble could it be?

. . .

Before Earl's call, my life was as close to normal as I would ever know.

I was thirty-three years old, the brand-new dad of a six-month-old baby boy we had named Teddy, and two months from making partner at Taft Stettinius & Hollister LLP, one of the oldest law firms in Cincinnati. My wife, Sarah, and I were getting ready to build a house more suitable for a family than the century-old house—a handyman's special

sold to someone not especially handy—I'd bought six years earlier, before
we were married. Sarah and I had met in the fall of 1991, a little over a
year after I'd started working at Taft. I was having coffee with two colleagues, another guy in his twenties like me and a young female colleague
named Maria, when I complained that working the long and unpredictable hours of a new associate at a big law firm made it hard to keep up
any kind of social life. That was especially true in Cincinnati, where
people tended to stay put their whole lives, keeping the friends they'd
had in high school through adulthood. "I have friends I can set you up
with," Maria said. She promptly called Sarah Barlage, twenty-five to my
twenty-six, who had just started as a lawyer for another corporate defense
firm downtown. Maria and Sarah had grown up and gone to school together in the same neighborhood across the Ohio River, in Northern
Kentucky. Maria described the two new single men at Taft, me and the
other guy. I was "the quiet one." Sarah was smart and lively and liked
life-of-the-party types. Based on the descriptions, Sarah had a definite
preference for bachelor number two. "But I'll take the first available."

I made sure that would be me. Maria and Sarah and I met for lunch.
My first thought upon being introduced to her was the typical response
of a twenty-six-year-old male. She was gorgeous. The two women, both
extroverts and champion chatters, talked the entire time as I listened. I'm
sure I was staring at Sarah, riveted by her beautiful bright blue eyes. She
was so full of life, she virtually glowed. She was charming and gregarious
and seemed to see everyone in their best light—everything I wished I
was but wasn't. I doubt I said more than a sentence or two, but I was
hoping that I emitted an air of appealing mystery. Maybe not.

After that lunch, Sarah wasn't exactly waiting breathlessly by the
phone for me to call, but at least she didn't turn me down when I asked
her to dinner and a movie. We sat through *Cape Fear*, and only later did
she confess that she actually didn't like going to movies. She says now
that during dinner I spoke of my sister and parents in a "tenderhearted"
way, and that had made her look past my awkwardness. It was a lot to
look past. How clueless could I be? The night that I decided I was going
to propose, she came down with a raging migraine. I stopped by to see

her, but she couldn't even raise her head off the bed. She was just lying there in the dark. It was all she could do to ask me to go get her some pain medicine. I decided that this was my chance and put the ring in the prescription bottle. I turned on the light so she could see the ring when she opened the bottle. She was like "I love it. Now turn off the light and go away." But somehow she could see through my blunders to my best possible self, a self I doubt would have been possible if she hadn't agreed to marry me.

She moved into my old house, but when she got pregnant we began looking around: at the walls (probably toxic lead paint, we decided); at the neighborhood (a lot of twentysomethings and almost no families); and at the appealing wooded, hilly, affordable neighborhoods on the other side of the Ohio River in Northern Kentucky, where Sarah had grown up.

We decided that my impending new job security and the higher income I'd have as a partner would not only allow us to build a suitable family home in those Kentucky hills but also allow Sarah, who defended corporations in workers' comp claim cases, to step back from the crazy hours and, for a few years at least, devote herself to family life.

Every morning I pulled my 1990 Toyota Celica into a parking deck filled with Benzes and Beemers, took an elevator up twenty floors, and walked down hallways decorated with contemporary art prints and oil paintings of long-dead partners. Those partners included the son of President William Howard Taft. Founded in 1898, Taft was, by the time I started working there, viewed as a hive of Ivy League grads with old money and impressive family pedigrees. As an air force brat who had moved half a dozen times before graduating from public school, I was still getting acclimated to that world of bow ties and seersucker suits, where not long ago it was considered gauche to walk down the hall without a jacket on and business cards were seen as tacky. During law school, I had applied for Taft's summer internship program. I hadn't made it past the first interview. But after I graduated from law school at Ohio State, somehow I made the cut. I'll never forget my first day in 1990, when an older partner greeted me with a knowing nod.

"Harvard?" he said.

"Nope," I said.

"Virginia?"

"Ohio State."

"Ah!" he said. "The dark horse!"

Eight years later, and I was still trying to shake off that feeling of being an outsider, hungry to prove—to myself as well as my partners—that I belonged at Taft. What I lacked in laurels I made up for with a high tolerance for tedious work. I didn't mind the painstaking tasks, the hours of immersion in minutiae, that most lawyers dreaded and avoided. I'd always been a little detail oriented to the point of being compulsive. As a kid, long before I could get a learner's permit, I had been fascinated by cars, vacuuming in every facet of model, make, and year. Like a walking auto encyclopedia, I could recite the fine-print differences between the '72 and '73 Buick Electra, for instance, and I was fond of those thousand-piece model car kits where assembling the headlamps required tweezers and a magnifying glass.

I never would have imagined that my obsessiveness would come in handy as an attorney, but it had not only proven useful but made me somewhat impervious to the boredom so many new associates suffer.

My character made me a good fit for the work of the law but not so much for the business of the law. Thus far, all of my work had been for other attorneys' clients. I didn't have my own book of business. In fact, I had never brought a new client to the firm, often one of the key indicators of success at a corporate firm such as Taft. I dreaded schmoozing the way most men dread shopping. I hated being that guy in a rented tux sipping white zinfandel at some charity gala, sweating over which fork to use. It went against my whole nature to compete with other lawyers trolling a crowd for leads. I was happy to let my colleagues do the wining and dining while I kept my head down and did their grunt work. I was what they called "a grinder, not a finder."

I worked in the firm's environmental practice group, a small contingent with a national practice and a fine reputation. Although that had been a bit of an accident. My last semester in law school, I had grabbed an elective course on environmental law only because the subject sounded refreshingly down to earth after the mind-numbing abstractions of contracts and tax law. Based on that and not much else, when I saw that Taft had a prominent environmental group, I asked to join it. I didn't have any real idea what the group actually did.

As I soon learned, most of our clients were large corporations, and much of my job involved dealing with permits, regulatory filings, and related lawsuits. Basically, I helped companies avoid breaking the many environmental laws and regulations governing their waste streams. Around the time Earl found me, I was busy helping our corporate clients with contaminated hazardous waste sites being cleaned up under the federal Superfund law.

When I called my grandmother to thank her for her referral—I left out the part where I had been rolling my eyes—she told me I had a closer connection to Earl than I realized. As a boy, back in 1976, I had visited the Graham farm, right next to the Tennants' land. I remembered that visit vividly, a happy memory. I had even milked one of the cows whose offspring might be drinking the foaming water flowing through the farms.

Two weeks later, I met Earl and his wife, Sandy Tennant, in Taft's glassed-in reception area. Earl, dressed in jeans and a plaid flannel shirt, rose from a midcentury modern couch. He towered over me. His face broke into a grateful smile, and he offered a callused hand. Sandy, dressed in church clothes, smiled shyly. I glanced at the four or five cardboard boxes they'd brought. Earl hadn't been kidding: they were loaded with documents, videotapes, and photographs.

We rode the elevator up to the Gamble Room, one of our larger conference rooms on the twentieth floor, the floor where the entire environmental practice group was encamped. On the way, we bumped into Tom Terp, one of the two partners in charge of our group and the one who had taught me how to do the Superfund work. Kim Burke was the other partner in our group at the time, the one who had taught me the complicated world of environmental permitting and compliance counseling. I had enormous respect for Tom and Kim. They were generous mentors who demonstrated and demanded an intense work ethic and impeccably high standards. If you got a late-afternoon call from a client with a question, you worked all night to have the answer ready first thing in the morning, whether the client needed it then or not. The point was to prove—over and over—that we did the best work and provided the best client service.

Having bumped into Tom, I invited him to join us and was relieved when he agreed. Earl was about as far from our typical client as you could get, and his claim against the kind of corporate client our firm usually defended could land us in ticklish territory. The prospect of having Tom's eyes on the initial meeting was immensely reassuring.

As I watched Earl and Sandy take in the surroundings, I imagined they could use some reassurance as well. The Gamble Room was a severe, windowless room. The off-white walls were decorated with modern art, including a signed Andy Warhol print that must have been worth more than Earl's entire farm. A framed photograph of Mr. Gamble, a long-dead partner, watched us from one of the walls, possibly with horror at the idea that we'd be considering a case like this. We gathered at one end of a gigantic dark wood conference table and settled into the blue upholstered chairs. As we did for all our client meetings, we had arranged a tray of soft drinks and a pitcher of water for our guests. But Earl and Sandy were not interested in small talk or refreshment. They wanted to get straight to business.

Earl launched into his story. His narration was a bit wheezy. Sometimes he had to stop after a sentence, as if his air had given out and he needed to catch his breath before continuing. The difficulty seemed to have no impact on him. He'd been trying to say what he had to say for a long time, and bad lungs weren't going to get in the way.

He had spent the last two or three years calling every agency he could think of. The local agencies had deflected or ignored him. He had then called the feds. The US Environmental Protection Agency (EPA) had claimed to be looking into the problem. At one point in 1997, some government types had showed up at his farm and spent a few days stalking around taking samples and writing notes, but he'd heard nothing back from them and seen no results. At the present rate, he said, his entire herd would be dead by the time they got him any useful answers.

"No one wants to get involved," he said.

He hoisted his boxes onto the conference table and started pulling out the things he had wanted to show me. We flipped through stacks of three-by-five photos of cows and wildlife—fish, frogs, deer—all dead. He had photographed a pipe spouting white foaming water into a creek. I could read a sign by the pipe marked with the E. I. du Pont de Nemours and Company logo and the word "outfall."

Out came the videos. Earl pulled out stacks of black VHS tapes with handwritten labels indicating "Dry Run" and "Lee Creek," with dates ranging over the last several years. An assistant wheeled in a cart with a nineteen-inch portable color television and videotape player. Earl handed me the first tape. I put it in and pressed play.

Grainy images, peppered with static, flickered on-screen. Earl had spliced his camcorder footage into a grisly highlight reel. Through the garbled audio, I could hear his angry voice-over explaining each scene as a parade of gruesome images stuttered across the screen.

By the end of the video, I was floored. This was powerful. I no longer had any doubt: something was terribly wrong here.

Earl's mood lightened when he sensed that someone was finally listening—and most important, that we got it. The fire in his eyes turned from anger to hope.

"Can you help us?"

• • •

That was a pretty big question.

After the Tennants left, I had a lot to consider. His outrage and passion had infected me: How was it that no one else—the other attorneys and government officials he'd contacted—had been able to recognize such an obvious problem? I'd seen *Erin Brockovich* and *A Civil Action*. I knew that the evil, faceless corporation could sometimes be real. But I had also worked with men and women in any number of corporate offices who were decent, moral people, and the last thing they'd have wanted to do would be to harm their fellow citizens. So I was reluctant to jump to conclusions and embrace Earl's narrative that everyone locally and "at the state" was working for (or with) DuPont or scared of the repercussions of crossing them.

There was another element that wasn't in those boxes, but it had become clear as I listened to Earl that day. Earl himself was getting sick. He said he had trouble breathing after going anywhere near the landfill or the creek, particularly when the "vapor clouds" from the aerator at one of the landfill's ponds wafted over his property. He was frightened for Sandy and his two young girls, Crystal and Amy. He had forbidden them to go near the creek they had once liked to walk along and splash in as kids. But what about the well water they drank? What about the air they breathed?

I wanted to help him, no question. I wasn't the type of lawyer who had grown up dreaming of being like Atticus Finch, a crusader defending the weak against the corrupt and powerful. In fact, I'd had no real idea of what working at a law firm would be like until I arrived at Taft. I'd merely wanted a professional career that could generate a good living.

But now I was here, and someone who desperately needed my help had been sitting in front of me pleading for assistance.

It wasn't a simple choice. Taking this case could have serious ramifications. Taft would be suing the kind of company it usually represented. That kind of news can spread through the industry like a grease fire. I didn't want to tarnish the firm's sterling reputation in the business community. Could taking the case damage important relationships, or scare away future clients? Was I really going to make this the first piece of business I brought to the firm?

After Earl and Sandy left, taking their boxes with them, Tom and I lingered in the conference room to debrief. A pioneer in environmental law, Tom had launched his career soon after EPA had been formed in 1970. He had been involved in the country's first Superfund cleanups and was now viewed as a national leader in the field. He had a far better understanding of the risks and how the move would be perceived—by our partners as well as the chemical industry.

However, Tom had been as convinced as I by the photos and the video that there was a real wrong being done. He was not scared off by the unorthodox nature of the situation. This was a small case—just one farmer and his cattle. It shouldn't make enough of a splash in the pond to scare away the big corporate fish our firm worked with.

Not to mention, it was the right thing to do.

Besides, he said, a little plaintiffs' work now and then could make us better defense lawyers.

There was one other problem. Tom pointed out right away that there was no way Earl's family could afford our hourly rates. Taft, like most big defense firms, billed several hundred dollars an hour. For many of the corporations we served, that was just another cost of doing business that would have a clear return on investment. It was our job to limit a client's liability, which could be enough to put a company out of business. The money we saved the company by avoiding penalties or defending lawsuits far exceeded our hourly fees.

Plaintiffs' firms, on the other hand, represented clients who often could not afford to pay hourly legal fees. Usually they worked for contingency fees, which meant they took no payment until (or unless) they negotiated a settlement or won a trial. Then they received a predetermined percentage—usually anywhere from 20 to 40 percent—of their client's award. It was an all-or-nothing gamble. Unlike defense lawyers, who were usually compensated regardless of the outcome, plaintiffs' lawyers were paid only if they prevailed. If they lost, they had to eat their costs. The firm would take a loss, but the plaintiff would be none the poorer.

The risk of not getting paid was not the firm's alone. The currency of defense firm practice is billable hours—time spent working on behalf of clients that can be charged to the client, supporting the financial well-being of the firm and justifying the individual lawyer's existence. At many large defense firms, an adequate performance was considered to be around 1,800 to 2,000 billable hours in a year. Lawyers who racked up much less than that risked not only getting no raise (or maybe even having their salary cut) but potentially derailing their entire careers if not corrected quickly.

So I swallowed a little hard when Tom agreed that Taft would represent the Tennants for a contingency fee and I would be the partner in charge of the case and the billings. It seemed fairly straightforward: I'd pull the permits, identify which chemicals were going into the landfill, and figure out what limits were being exceeded, just as I'd done so successfully in my corporate defense work. I would simply add this case in with all the other cases I was already working on for Tom and Kim. Our assumption was that the Tennants' suit wouldn't take very long or demand too much of my time, and it would be a nice chance to cultivate a client of my own on matters I could handle myself. I suspect that Tom was thinking it wouldn't be a big gamble of firm resources, since there was a slim chance I'd otherwise be bringing in any big new corporate accounts. Maybe he saw it as a way to coax me out of my shell.

3

PARKERSBURG

June 8, 1999
Parkersburg, West Virginia

Eight months after Earl's visit, I put on my meet-the-client suit and slid behind the wheel of my little Celica for a drive to West Virginia. I wanted an up-close look at the physical environment, hoping it would yield some obvious clue that would lead me to the source of the problems. So far, I hadn't found much.

After I'd spent all those months on the case, that was a damning admission. I'd been sending Freedom of Information Act requests to all the relevant agencies, poring through the thousands of records we had received in response relating to the permitting and operation of the landfill—all as I continued to work on Superfund cases with Tom and regulatory work with Kim. As required, DuPont had been sending the government regular reports on tests on the landfill water outflow, but the reports showed nothing that seemed to explain what Earl was seeing in his cows. I'd go to our environmental library—a grandiose label for a closet-sized room on the twentieth floor filled with environmental reference books—and browse through the volumes on various chemicals, looking for any that might cause the foam Earl had found in Dry Run Creek. I wasn't finding any matches.

In all the work I'd done for the firm, I'd always been able to dig and find the needle in the haystack, which was why Tom and Kim had given me sufficiently positive reviews to justify naming me a partner, even if I'd brought in none of my own clients yet. For the first time, and with the first client I had brought to the firm, no needles were appearing. So I dug harder, adding hours to my already long days. I usually came in at around 8:30 six days a week and stayed at least until 6:00 or 6:30. Now I was staying until 10:00 or 11:00.

. . .

As the days went by, Earl called several times a week wanting to know what I'd found and how much I'd figured out. I could sense his growing frustration when the answer was always the same, that I was still hunting through documents. Let's wait and see what the rest of the government records say, I'd tell him. This hardly satisfied him. For years now, he'd been dealing with "those people"—the state, EPA, and most of all Du-Pont—and he didn't trust a word out of their mouths. I understood how he felt, but I relied on the science: samples would be taken, tests run, and any problem would be revealed.

The calls all ended the same way: Earl would hang up, clearly unhappy—both with the progress of the case and with me personally. I had never had an unhappy client before, and it weighed on me.

Having found nothing useful through ordinary channels, I wrote up a formal complaint to initiate a lawsuit against DuPont, which would allow us to begin the legal discovery process, the only way I could think of to pinpoint the problem chemicals. Only DuPont knew exactly what chemicals were in Dry Run Landfill, and they were likely the only ones who knew the toxic effects of those substances. DuPont did extensive in-house toxicology testing, which wasn't rare. Big chemical companies like DuPont often knew more about the toxic effects of their chemicals than the regulators did.

Presenting my legal plan to the Tennants would be a good opportunity to meet the entire family, tour the farm, and show the flag, reminding them that I really was on their side. Once they signed the formal paperwork, I would file the complaint with the court, and the lawsuit would officially begin.

As I left Cincinnati and headed east, the freeways shrank into four-lane roads that wound through the wooded hills of southern Ohio. It was surreal, driving along the same roads I had traveled in the back seat of my parents' 1967 Lincoln. I hadn't been to Parkersburg since 1985, when Grammer had moved to live closer to my parents in Dayton. I planned to drive by Grammer's and my great-grandparents' houses and visit a few childhood landmarks. It felt a little like going home.

. . .

I drove through downtown Parkersburg, surprised at how withered it looked. Dils department store, where my sister and I had ordered egg salad sandwiches at the cafeteria, was long gone. The historic brick buildings looked tired and run down. The streets felt empty. Many of the storefront windows were boarded over. From what I could tell, the businesses that had once thrived downtown had shut down or moved closer to the mall in nearby Vienna.

Parkersburg had been settled shortly after the American Revolution, but it had come of age in the era when rivers were our freeways. In the nineteenth century, flatboats, keelboats, and steamboats ferried people and cargo up and down the river between Pittsburgh and Cincinnati. A natural stopping and resupply point, Parkersburg flourished as a result of the traffic. As the boom in rail construction in the latter half of the nineteenth century caused trains to replace river barges for shipping goods, Parkersburg declined. But as one industry withered, another was about to explode.

In 1859, prospectors drilling for brine (to make salt, a precious commodity) struck oil. They kept drilling. Each new well seemed more generous than the last. Soon they were pumping 1,200 barrels a day out of West Virginia. Parkersburg, serving as a hub and supplier of oil field equipment and goods, saw a second boom. Just when it looked as though things couldn't get any better, the area was found to be rich in natural gas.

The oil and gas money transformed the roughneck river town into a cultural oasis. Vaudeville shows drew crowds to Parkersburg's 1910 Hippodrome. Couples danced to Benny Goodman and other live bands at the Coliseum, West Virginia's largest ballroom. Horse-drawn streetcars gave way to electric trolleys that ran every fifteen minutes, day and night.

By 1900, though, the oil fields began drying up, and with them the local economy. In the 1930s, the Great Depression shuttered many of Parkersburg's banks and businesses. One of the companies that kept the city afloat through these dark years was the American Viscose Corporation, among the world's largest producers of rayon, the first man-made fiber. Rayon, created as an artificial substitute for silk, was made through the chemical manipulation of plant cellulose. Employing five thousand people at its peak, the American Viscose plant was a source of economic stability and local pride. Its employees, fiercely loyal, thought of the company as a family.

American Viscose also introduced the area to the chemical industry. West Virginia was viewed by many as a chemical company's dream. Though the oil reserves were tapped out, there was still plenty of coal to fuel factories. Cheap labor was abundant. The hills held huge reserves of salt brine, a necessary ingredient in many chemical processes. And the rivers provided an ample supply of water for shipping, industrial cooling, and liquid-waste disposal. As the twentieth century soldiered on, the Kanawha River Valley, a hundred miles south of Parkersburg, soon had the highest concentration of chemical companies in the world. It came to be known as Chemical Valley.

As World War II was ending, DuPont bought a swath of bottomland that had originally been awarded to George Washington following the Revolution. It was south of town on the eastern bank of the Ohio River, a fine place to build a new plastics plant. DuPont (everyone in town stressed the first syllable: DEW-pont) quickly became the economic wellspring of the town. Everybody knew somebody—somebody lucky—who worked at the DuPont plant. DuPonters were in a social class of their own, one rung above middle class.

Parkersburg residents saw chemical-manufacturing jobs as first-rate work. In a town where the median household income was thousands of dollars lower than the already depressed state median, chemical workers often earned twice that amount. And DuPont had deep pockets for local giving—to schools and nonprofits and for civic projects such as parks—so the whole community benefited, not just the employees. To the people of Parkersburg, the plant meant jobs, economic stability, and a sense of civic security. It meant hope.

. . .

For lunch I stopped at a local diner, where I met a fellow attorney. I was licensed to practice law in Ohio, but to file a case in West Virginia courts, I would need local co-counsel. Larry Winter, a corporate defense lawyer like me, was a quiet man with closely cut hair, a soft voice, and a gentle disposition. We had met while working side by side at a West Virginia Superfund site where Larry had served as our corporate client's local, West Virginia counsel. Back then, Larry had been a senior partner at Spilman Thomas & Battle, a prestigious Charleston, West Virginia, defense firm with clients similar to Taft's. He had since left Spilman to start his own small firm, also now based in Charleston.

When I first told Larry I had taken the Tennant case, he was shocked. To the extent it's even possible, Larry is more understated than I, so I didn't realize how truly taken aback he was. He later admitted he'd found it "inconceivable" that Taft would allow me to sue DuPont. But after hearing about what was happening on the Tennants' farm, he agreed to help me out, even if it meant jumping the fence to the "plaintiffs' side."

After lunch, Larry got into my car and we drove out to the Tennants' place together. The farm was located in the hills south of town. On the way there, we had to drive past the DuPont plant—it was called Washington Works in honor of the land's original owner. Looming along the river like a city in itself, a fortress of industrial buildings knit together by a web of metal pipes, the plant's towering stacks belched columns of smoke. As we drove along DuPont Road, a two-lane byway that ran adjacent to the plant, I studied it through the window. It just kept going . . . and going . . . and going. It had to be almost a mile long.

A plant that size generated tons of industrial waste—literally tons—every year. It came in all sorts of forms: airborne particles wafting from the stacks; industrial sludge in every viscosity, some dumped in on-site sludge pits and other, more liquid waste discharged into the river. The company was required to obtain permits that restricted what and how much could be discharged and where.

Some industrial by-products had to be shipped to other DuPont facilities for treatment or disposal in special landfills. But many of the solids were trucked to local dumps.

The permits confirmed what the Tennants had alleged: that the Dry Run Landfill adjacent to the Tennants' land, owned and used exclusively by DuPont to dump plant waste, was limited to construction dirt, railroad ties, office trash, and other nonhazardous waste. I had discovered in my research that some regulated contaminants—arsenic, lead, and heavy metals—were also being monitored for there. Yet they were not the kind of pollutants that should cause the types of health issues—black teeth, tumors, wasting—plaguing Earl's animals. Also, none of them seemed to be a candidate to make the water foam.

Regardless, I knew from my work negotiating landfill permits for our corporate clients that the unauthorized presence of regulated or hazardous materials in a landfill that wasn't built (or permitted) for their disposal could generate liability for DuPont. It seemed from the government files I'd reviewed that Dry Run Landfill lacked adequate features to prevent dangerous chemicals from overflowing or leaching into the surrounding environment during heavy rains. What's more, it appeared that industrial sludge had been dumped into the landfill, but it was not clear to me whether DuPont had received proper permission to begin dumping that waste in a landfill designed for nonhazardous waste. I thought those discrepancies might possibly form the basis of our suit.

Larry read me Earl's directions, which I'd hastily scribbled on my legal pad. They took us off the main road and into the hills and hollows. The farther we got from the plant, the more the asphalt narrowed, climbed, and twisted. Just when I thought we were lost, we spotted the landmark that matched Earl's description: an old mailbox on a post by the road. Just off the road, a small log cabin lay empty. Farther down the unpaved drive, a new house was under construction. It had recently rained, and the rich scent of damp earth poured in through the windows. We turned onto the long, narrow driveway, gravel crunching and popping beneath us, and stopped when the gravel gave out and our wheels sank into a soggy pit.

Larry and I straightened our ties, cleared our throats, and prepared to meet the clients. I swung open the car door and set my left wingtip into the mud.

4

THE FARM

June 8, 1999
Parkersburg, West Virginia

As soon as my dress shoe squished into the West Virginia mud, I felt a sudden wave of self-consciousness. I was wearing what I always wore to client meetings: dark suit, white button-down shirt, conservative tie. Larry wore the same uniform. It hit me only then how clueless my choice of apparel had been. What had I been thinking? These people would take one look at us and decide, reasonably, that we must be idiots.

A large dog came running to greet us, jumping up on my suit with muddy paws, wagging his tail as though we were his new best friends. Earl strode out of the house in denim overalls and heavy work boots, charitably ignoring our outfits.

"Hey, howdy!" he said, wiping his hands on his overalls before greeting us with a firm handshake. He looked genuinely happy to see us. Again I was impressed with his imposing size and air of country toughness. I also noticed that he was wheezing even more than he had been when we'd met in Cincinnati, and he looked worn and tired. He could barely get through a sentence without stopping to grab a breath. Earl waved us into the unfinished house, which he was building board by board. The main floor was framed and bare, awaiting drywall. We followed Earl downstairs to the finished basement, where he and Sandy were living while he finished the house around them. Sandy was sitting at the kitchen table with a couple who looked less happy to see us. Earl introduced them as his brother and sister-in-law, Jim and Della Tennant. Thinner and slighter than Earl, Jim nodded curtly from under his ball cap. Della, not quite five feet tall, kept her arms crossed and smirked at our suits. She wore an expression I see a lot on people who don't trust lawyers.

We sat around the kitchen table, where Earl pulled out scrapbooks and videos we hadn't seen at Taft. He held jars of discolored water up to the light. Assorted animal bones were passed around the table like show-and-tell, including a cow skull with blackened teeth. He dug through his freezer and pulled out stiff samples of organs and tissues, all collected during his "autopsies" in the field. We flipped through stacks of three-by-five photographs of sick and dead animals. This was way more evidence than I had expected. It was more than impressive; it was humbling. I was used to wading into Superfund cases alongside other lawyers who might argue passionately about who was responsible. Serious issues and large dollar amounts were at stake. But those cases could proceed with professional detachment; ultimately, they were the company's problems— not the lawyers'. At the end of the day, they'd return to their comfortable homes far from the threat of the chemicals, knock back a cocktail, and forget about their work until the next morning.

The Tennants had no respite. They were living the case day and night, each morning wondering if they'd wake up sicker than they had been the day before or find another of their animals dead on the ground. For them, there was no safe distance.

As the painstakingly gathered evidence was presented, Sandy listened quietly, watching her husband. The reticence Jim and Della had displayed on our arrival disappeared. They jumped in with rapid-fire stories: about finding dead deer lying in the fields; about wheezing and choking on the air near "the fill"; about burying dogs, cats, and other pets that had all died mysteriously.

I struggled to keep up with the barrage of information. My world revolved around order and protocol. I was used to meeting agendas, structured interviews, and formal procedures. The people I worked with were trained to convey information in a certain manner, sequence, and pace. Here, the details and stories were pouring out in great volume and in no particular order.

At first I tried to take organized notes. There was so much to process: documents, videos, photographs, jars of funky water, frozen cow livers, wild-animal bones. So many stories. And stories within stories. All of it needed to be collected, reviewed, analyzed, and organized.

But it was like drinking from the proverbial fire hose. Soon I shoved my legal pad aside and just listened. The talk went on for hours. At some point, Earl and Jim's brother, Jack, walked in and joined us. Less animated than Earl and Jim, Jack, a metalworker, was shorter and squatter than either of them and clearly less keen about this whole deal with lawyers and lawsuits. I'm guessing that part of the reason was that Jack's livelihood did not depend on the farm. He sat quietly and listened, occasionally jumping in to add a detail or answer a question, but he seemed as though he wasn't yet sure if he wanted to get too involved. Here were three brothers with three distinct personalities. They disagreed on a lot of things, but they shared one certainty: Dry Run Landfill was the problem.

· · ·

Sometime in the early 1980s, DuPont began buying pieces of land from local families, including the Tennants. This wasn't at all uncommon. When factories needed land to expand, they'd go knocking on doors of local landowners with offers that could be handsome. Most residents were more than willing to sell. They already thought of local companies as good neighbors. Besides, when the factories grew, it meant more jobs and a stronger economy that seemed to benefit everyone.

Over the decades, the Tennants were one of those families. Different family members had sold different tracts of land to DuPont at different times. Decades before, Earl, Jim, and Jack's mother, Lydia, had sold eighty-some acres when she had been offered a deal she couldn't refuse: DuPont would purchase and own the land, but she could keep using it.

Through an annual lease for a trivial, token payment, her family was ensured the grazing rights for two generations.

So it wasn't unusual when DuPont approached Jim Tennant around 1980 about buying a few acres. Jim had worked on and off since 1964, doing manual labor at the plant. Jim and Della didn't want to sell. But DuPont persisted. Turned out that those few acres were smack dab in the middle of the land they envisioned for a new landfill. Every few years, the folks from DuPont would stop by and try again to persuade them. Jim and Della were assured—verbally and in writing—that no hazardous chemicals would ever be disposed of in Dry Run Landfill. "The landfill will contain nonhazardous wastes only, including ash from plant boilers, waste plastics, glass, scrap metal, paper, and trash, all transported by covered truck or closed containers," said the landfill information sheet to be shared with landowners. "It will be designed and operated with full consideration for neighbors. The ash will be wetted, and the road will be paved to control dust and mud. The fill will be covered with dirt at the end of each day. Daytime operation is planned. The site could receive ten to fourteen truckloads per day."

As Lydia had been promised years before, DuPont assured Jim and Della that they'd retain the right to graze their cattle along Dry Run Creek. Those conditions finally convinced the Tennants to sell. They didn't really want to let go of their land, but if they could both sell it and continue to use it, the deal made sense and they could always use the extra money. They sold a few acres, which DuPont patched together with plots of land they'd bought from several other families until they owned a contiguous seventy-seven acres that made up a little valley. That valley became Dry Run Landfill.

Once operations began in 1984, Jim and Della watched dump trucks roar past, delivering load after load into the valley, slowly filling it up.

. . .

I was eager to see the land and the creek that backed up to Dry Run Landfill. Larry and I ditched our jackets, pulled off our ties, and rolled up our sleeves for a farm tour on a ninety-degree summer day.

The Tennants took us first to "the home place," the little farmhouse where Lydia had raised them. We stood next to a small creek at the bottom of a sweeping hillside, watching cattle grazing on impossibly green grass. We saw some skinny, sad-looking cows that were among the last of his once abundant herd, but we didn't see any foam in the water; Earl explained that it was too far out on the property for us to walk to. We'd seen it all in the video anyway, and it became clear that what he really wanted was to show off his farm. Even after all that had gone wrong, his pride in the place was palpable. The Tennants showed us the barn and the grain silo they had built by hand with concrete blocks that fit together like Lincoln Logs. It held enough corn and silage to feed a large herd all winter. I could smell the hay accumulating in the barn's giant hayloft, which looked as big as a basketball court. As they cut and dried summer grass in batches, it would fill to the twenty-foot ceiling.

The sights and smells triggered memories of the Graham farm, not far from where we stood. The smell of cut grass, the lowing of cattle, the way the animals trusted and nuzzled Earl stirred something primal and compelling in me. I had always loved animals, probably beginning with the deep emotional connection I'd had with a black-and-white dachshund–fox terrier mix my family had adopted in 1963. Charlie, a constant companion throughout my childhood, had died on Christmas Eve 1981 when he got sick while we were visiting Grammer in Parkersburg. It was my one unhappy memory associated with the town—until I visited Earl and saw what he had to show me. I thought I understood Earl's connection to his cattle. To him, they were like family members, not livestock. He felt a duty to protect them. Now, I realized, so did I.

This was no longer some abstract "issue." It was a living, breathing problem with an entire family's future in the balance. They were trusting me to fix it.

· · ·

Three days after our trip to the farm, I finished the Tennants' legal paperwork and sent it to Larry Winter, who filed the complaint in federal court in West Virginia. In it, we alleged that DuPont had improperly managed waste at the Dry Run Landfill, violated their permits, and released harmful materials from the landfill. If our suit ultimately prevailed, we could force DuPont to clean up the landfill, and hopefully that would remove whatever was contaminating Earl's farm and poisoning his herd. We also included a negligence claim for property damage—to both the land and the cattle—and a personal injury claim for Earl's and his family's health problems.

The filing was mailed to the court and also to an agent, who would formally serve notice on DuPont. Within a day or two, a judge would be assigned. I made Earl understand that we could be in for a long process of legal maneuvering before the judge would even assign a court date. Almost nothing in the legal process happens quickly. But at least the lawsuit was official.

Now I had time to think and worry. Would DuPont's lawyers be shocked? Outraged? Would they see me as a traitor? Plaintiffs' lawyers and defense attorneys tend to view each other as totally different breeds, each world having its own culture, habits, and personalities. I didn't know a single attorney (other than Larry and I) who had ever "changed sides." It was hard to know what to expect. It was hard not to worry.

Five days later, my office phone rang. It was DuPont's in-house counsel. I braced myself for a harsh encounter. But the voice on the other line was fairly neutral. And it sounded strangely familiar. As soon as I placed it, all my tension evaporated.

"Bernie!"

Bernard Reilly was a lawyer I knew and trusted. We had worked together before. If Bernie was shocked by a lawsuit from Taft, his voice didn't show it. Calm and collected, even friendly, he was chatting with me like a colleague. He was someone who knew the rules and understood the seriousness of breaching them.

I had met Bernie through the same Superfund case where I'd gotten to know Larry Winter. I'd spent hours at the negotiating table with Bernie and other attorneys, figuring out how much each company that was suspected of sending or creating waste at the site should pay for its cleanup. A room full of lawyers duking it out over money can get considerably heated, but Bernie and I had always treated each other with mutual respect.

More senior than I, he had the slight frame of a guy who woke up at 4:30 a.m. to go for a run with his dog before work. He struck me as intelligent and responsible, a hard worker with sound professional ethics. He was someone I felt I could count on to remain cordial even if we sat on opposite sides of the table. Having Bernie as my DuPont counterpart seemed like a stroke of luck.

After exchanging brief pleasantries, Bernie delivered what he framed as good news: there was already an investigation of the Tennant farm under way. He didn't tell me then, but I found out later, that the investigation had been prompted by Earl's complaints to EPA. With EPA officials buzzing in DuPont's ear, the company proposed a joint investigation.

Partnering with EPA, DuPont had created a "Cattle Team" of six nationally recognized veterinarians—three chosen by DuPont, three by EPA. The extensive (and expensive) study aimed to identify the source of the cattle's problems and served to get DuPont out ahead of a potential problem in a way that would still give them some control of the process.

Bernie said that the Cattle Team vets had been working for several months now and their final report was due in as soon as a couple of weeks. He promised to send me a copy the minute it landed on his desk. He suggested, in the meantime, that we hold off on full-blown discovery— the enormously labor-intensive first phase of a lawsuit, when both sides request documents and information. Why waste time and effort finding out what the report would eventually tell us? While we waited, he said, DuPont would send me their internal documents relating to permits, deeds, and regulatory filings for the landfill. Through my prior Freedom of Information Act requests, I already had most of the documents that were available in the public files, but now I would get to see what additional materials DuPont had that could shed more light on what might be affecting Earl's cattle. In turn, I'd send DuPont copies of the photos, videos, and other evidence supporting the complaint.

This all seemed perfectly reasonable. I'd get the same types of records I often analyzed for my corporate clients; Bernie could have a better look at the problems on the farm. By month's end, we should have the Cattle Team report, which would tell us which specific substances to focus on, and we'd move forward from there.

This was in June 1999, nine months after Earl first dialed my number. One week became two. Two weeks became four. Two months passed. Three months. Then four. Every week, it seemed, I was told that the Cattle Team was "almost done." Summer became fall, and fall turned into winter. A couple of weeks before Thanksgiving, Sarah gave birth to our second son, Charlie. The delivery was fast and smooth (easy for me to say) and, thank goodness, without complications. I was thrilled for Sarah and myself, of course, but also for little Teddy, now just over a year old, who would have the little brother I'd always dreamed of having. By Christmas, while learning the strange math that made two infants ten times as hard to keep up with as one, I was still waiting on the Cattle Team. What on earth could be taking so long?

I started to feel suspicious. By this point, the federal judge overseeing the case we had filed for the Tennants had already scheduled the trial for mid-2000. That was now just six months away, and I hadn't even been able to start fully preparing my case. Every time Tom Terp asked me about my progress on the Tennant case, I felt a knot tighten in my gut. Nobody in my firm said anything indicating concern about the fact that the case was stretching on far longer than anticipated or the unremunerated hours I was spending with no financial reward in sight. But they didn't have to. I began to get a queasy feeling, a self-consciousness about my lack of production. To make things worse, the case, which was supposed to be straightforward, familiar ground for me, was only getting more complicated.

During that six months of limbo waiting for the cattle report, in addition to all my typical corporate defense projects for Tom and Kim, I spent countless hours poring over the paperwork for the Dry Run Landfill that I'd received from DuPont in lieu of full discovery. I reviewed permits, property deeds, and regulatory filings, along with the records I'd been receiving through my Freedom of Information Act requests from EPA and its state counterpart, the West Virginia Department of

Environmental Protection. Scouring all the paperwork from the government and DuPont, I thought I'd finally identified all the waste materials monitored at the landfill, but still, nothing was jumping out as an obvious cause of the problems on the farm.

Though the case had slowed to a crawl, time did not stand still on the farm. Earl called me almost every week with news of more dying animals. With each call, I could sense his frustration and anger rising. I couldn't blame him. It was starting to feel like a hostage situation—every passing day without relief might claim another life.

The first week of the new millennium, six months after Bernie Reilly and I had first spoken, the Cattle Team report landed on my desk in a stack of mail. The report stood out: a thick stack of photocopied pages with a hefty appendix of photos and data, obviously built to impress. Elated that the answers that had eluded me might be sitting in that stack of paper, I dug in.

As I flipped through the report, my optimism evaporated. It failed to identify a single chemical released from the landfill that could be linked to the cattle problems. The six veterinarians claimed to have thoroughly examined Earl's animals—drawing blood, taking tissue samples, conducting all sorts of tests. They had analyzed the pasture where the cattle grazed, the water they drank, and their supplemental feed. According to the report, they had found . . . nothing. (At least nothing associated with the landfill.) Their conclusion: *There was no evidence of toxicity associated with chemical contamination of the environment.*

I read on with growing dismay. The report's conclusions were not just surprising; they were insulting. The cattle's poor weight gain was blamed on "fly worry." The "manic stampeding of the cattle" was "likely due to the large number of flies."

The poor conception rate was due to "a lack of intensive breeding management" by the farmers. The depressed calf size was blamed on "lack of a balanced diet." The report attributed the cattle's problems to malnutrition, pinkeye, copper deficiency, and endophyte toxicity, a sort of poisoning by a fungus that grows naturally in fescue grass. The clinical, laboratory, and historical data, according to the report, supported the team's conclusion that "these four conditions can readily account for the chronic herd problems on the Tennant farm."

What about the dying wildlife? The twenty cases of other dead animals presented in the videotapes were declared to be "of no diagnostic value." The Cattle Team considered all of those deaths to be "incidental" because "there did not appear to be any consistency in the species, location, or timing of the deaths."

· · ·

The Cattle Team blamed "deficiencies in herd management, including poor nutrition, inadequate veterinary care, and lack of fly control." Its recommendation was "for the owner of the Tennant herd to engage veterinary and nutritional consultants in the design of a herd health program."

In other words, it was all Earl's fault. According to these "experts," every single one of the 150 cattle deaths on the farm was a result of in-adequate animal husbandry. Earl simply must not know what he was doing. Or, worse, he was mistreating his animals.

My stomach sank. I let the thick package drop with a thud and pushed away from my desk. I was too upset to sit. I paced around my office as my mind raced. The conclusions blaming Earl for the devastation to his herd were overkill at best—too certain and too damning, painting a picture of a situation that bore no resemblance to what I'd observed on my visit to the farm.

Worse, I wondered if *I* had screwed up. It hit me all at once. I now saw that I had been beyond naive to think it was just lucky coincidence that Bernie Reilly had been assigned as my opposing counsel. I had to consider the possibility that he'd been chosen specifically. Had DuPont wanted to use our prior relationship and mutual respect in past dealings to lull me into letting down my guard? I now feared that that was exactly what had happened. I was furious.

When Bernie had suggested we delay full-blown discovery since there was a cattle report only a couple of weeks from completion that might answer most of the questions, or at least narrow the possibilities down considerably, I'd thought he was being helpful. I now wondered if it had been a cunning trap; whether DuPont had been keenly aware a new partner at a corporate defense firm taking on the firm's first-ever plaintiff case would be under the microscope and sweating the costs. Now I'd learned the even steeper cost of having waited. The cattle report, with its EPA "experts," had answered no questions, narrowed nothing down, and left me with additional discovery as our only hope of finding out what was killing Earl's cows. Discovery was a time-eating monster, and I was almost out of time. I felt so ashamed of how easily I had permitted DuPont to delay full-blown discovery for so long.

But there was no time for self-pity; I had to push the shame aside and focus. Maybe what little time I had would be enough. It would have to be. Earl and his family were at risk, and they were depending on me to fix the problem.

. . .

If there was a bright side to the situation, it was that I now knew I would never, ever let my opponent lure me into complacency again.

It seemed to me that they sure were going to an awful lot of trouble to keep me at bay. Why? I'd initially been skeptical of Earl's conspiracy-theory attitude, but now I was just as certain as Earl was that someone was eager to keep something from us.

I wasn't sure what it was they didn't want us to see, but I was now hell-bent on finding it and bringing it to light. But first, I had a task I dreaded: telling Earl what the report said.

When I called him, I thought he might jump through the phone.

"This report is garbage!" he said, livid. The whole Cattle Team was "a total joke."

Earl reminded me that he hadn't trusted the competency of the vets from the outset. He told me they were arrogant know-it-alls—"eggheads"—who wore their PhDs on the sleeves of their fancy dress shirts and treated him like a dumb hick, telling him how to raise his cows. According to Earl, they'd paid little attention to the evidence he'd gathered but seemed to have no idea what to look for. How dare they imply he was killing his own cattle? The report cemented Earl's conviction that neither DuPont nor EPA could be trusted. He felt certain, even more than before, that DuPont and "the government" were working hand in hand to make him "just shut up and go away."

Earl had been successfully breeding cattle for decades. He had bought quality stock—registered Polled Herefords. He had "upgraded" the herd with top-shelf bulls. At the Ohio State Fair auction in the late 1980s, he had placed the winning bid on the second grand-champion bull to come out of the state of Ohio.

What had been happening on his farm wasn't Earl's fault. But how could I prove it? How could I possibly link who-knew-what in that landfill to the Tennants' dying animals?

"What are you going to do about all this?" Earl demanded.

I explained to Earl that, like him, I didn't scare off easily and I was more determined than ever to find out what it was that DuPont seemed so eager to keep hidden. I also admitted that with our trial approaching fast, I didn't yet have any answers. There was one thing I *did* know, though: I was done being Mr. Nice Guy.

The cattle report had ended any chance that the situation could be simply and quickly resolved. Now that the "experts" had come down squarely on DuPont's side, we were going to have to find and bring in our own experts, which at best would be extremely expensive and at worst ruinously so. We really had nothing specific for them to focus on, nothing to tell them to look for. They'd be starting essentially from scratch, trying to find some phantom menace the experts from the government and DuPont had entirely missed or were trying to hide.

As determined as I was, I also knew I'd put not only Earl but my firm into a perilous position. The stakes of the bet I'd placed on the case were spiraling higher and the chances of it paying off growing dimmer. I wasn't looking forward to briefing Tom and Kim on the cattle report. When I did, they understood what it would mean in terms of sinking more firm resources into a case some people were undoubtedly now wondering why we had ever agreed to take in the first place, but they never gave any hint of disapproval or regret.

I was lucky to have their support. And unlike most plaintiffs' lawyers, many of whom faced bankruptcy if their contingency cases drove them too far into debt, I had a salary to fall back on, and the firm had plenty of other clients who were paying by the hour to help pay that salary. Still, I was under no illusion that the outcome of the case was not critical to my professional future. At that time, our firm had two tiers of partners. I was in the first tier, still on salary and not yet invited to share in the firm's profits. Making the second, far more lucrative tier, often was seen to be a matter of time and the amount of new business you brought to the firm. The principals did not have to say it out loud for me to understand that the clock on my advancement had likely stopped and wouldn't start again until we all saw how the Tennant business resolved.

I worried that I might be letting down not only the Tennants and my firm but also my family. I worried about Sarah, who was bearing the labor and stress of managing a household with two young children virtually single-handedly. Thankfully, our two boys were easy babies who almost from birth slept soundly through the night. At this point, it was more than I could say for myself.

. . .

With so much time already wasted, Larry Winter and I waded into the new round of discovery with a vengeance. If DuPont was trying to hide something, we would try through more aggressive discovery to force them to reveal what that was. By this time, Bernie Reilly had farmed out the document production process to an outside legal firm Larry and I knew well: Spilman. Even if I had wanted to unload on Bernie for pulling one over on me, he was no longer my primary contact; plus I didn't want DuPont to know I'd shifted gears or telegraph that I was onto them.

Larry had made it all the way up the ranks to managing partner at Spilman before eventually leaving to start his own practice. We agreed that this was a positive development, since he still had a good working relationship with his former colleagues.

At first, our discovery exchanges were brief and cordial as Spilman produced the additional landfill permitting and regulatory documents we requested. Spilman even agreed to a joint request to push the trial date back to January 2001 to allow time for the additional exchange of discovery. Then, in the spring of 2000, the dynamic abruptly changed.

It occurred to me that since we were coming up empty on the landfill records and knew damn well that something was killing those cows, we ought to expand our search beyond landfill records to those of the plant that produced the stuff that was dumped into the landfill. When we requested the files from Washington Works, DuPont started pushing back aggressively. My new requests were dismissed as "overbroad and unduly burdensome," an "unnecessary expansion" of the case into "irrelevant" issues. I wasn't exactly sure why I had touched such a nerve, but the heated response only made me more eager to see what was in those files.

DuPont fired back by piling on the number and intrusiveness of their requests for personal information on the Tennants; tax records, health records, and even orders for inspectors to examine their properties and homes. Since we were alleging that their property values and Earl's health had all been compromised, DuPont had every right to do so, but I wondered if it was an attempt at intimidation. The company seemed to be sending a message: You keep coming after us, we'll start coming after you. This won't be pleasant.

I was beyond exchanging pleasantries with DuPont's lawyers. I was done worrying whether I was making my opponent uncomfortable. I was exhaustive in my requests. I had to frame my search broadly enough to pull in whatever might be out there that had fallen outside the scope of the initial, more "gentlemanly" landfill-specific discovery that DuPont had been so willing to exchange with me.

I began to feel that DuPont was trying to run out the clock before the trial. I sent what I considered to be perfectly ordinary requests for documents and questions ("interrogatories" in legal lingo) relating to what might have found its way into the landfill from the plant. Once they were received, DuPont's counsel typically delayed responding until the last minute of the thirty-day response time allowed under the federal court rules. Even then, their "answer" was often nothing more than an objection to the request.

In response, consistent with normal discovery protocol, I would send a strongly worded letter, essentially saying, your objections are meritless and you need to respond *now*. Then I had to wait again for their response to my response, and if they still refused, my only recourse was to file a "motion to compel" with the court, asking the judge to order them to provide the requested answers or documents. Once the motion was filed, I had to wait once more for them to file a response to my motion and then again for the court to rule. Time flooded down the drain.

This process went on for weeks and months. The longer it dragged out and the more DuPont fought, the more I believed what Earl had said from the beginning: that DuPont knew they had poisoned Earl's cattle and were trying to cover it up.

In my Superfund work I had come across bad actors in litigation, parties acting in bad faith. They were usually unsophisticated business folks who had gotten in over their heads and didn't know or fully understand the rules. DuPont was anything but unsophisticated. They knew the rules as well as I did, which made it all the more shocking that I now wondered if Earl was right—were they intentionally breaking them? Part of me still wanted to think that the explanation for DuPont's behavior could be more benign. Perhaps they'd gotten their hackles up when, instead of folding in defeat after the cattle report had placed the blame on Earl, we came back swinging, redoubling our efforts, asking to dig into the plant's files. Now they would have to expend large amounts of time and money in an extended legal battle they probably felt they should have already won, and their strong opposition to our current discovery requests might merely indicate how unhappy they were about it.

Around this time, Tom Terp wandered into my office, looking bemused. "You'll never guess who just called me," he said.

"Who?" I asked.

"Bernie Reilly."

"What did he want?"

"The gist was that my new partner needs to back off with all this unnecessary discovery."

My jaw dropped open. Bernie's trying to block me by going over my head, crying to my boss, was unusual, in my experience. My astonishment quickly turned to elation. I beamed a smile at Tom. He understood, as I did, that the call signaled that I was making progress—I was getting close to whatever it was DuPont might be hiding. Bernie's call smelled of something more than irritation. Something closer to fear.

If Bernie had been hoping the call would slow me down, it backfired; Tom and I didn't quite exchange high fives—there were no high fives at Taft—but close enough. It was a great moment in another way. If I'd had any worries that Tom might be looking for an excuse to back out of this increasingly messy case, his reaction more than reassured me. He backed me up without hesitation. No one told Taft how to represent our clients— especially a lawyer from the other side.

. . .

So on it went. The next few months turned into a stalemate. I'd ask for records from the plant. DuPont would object. I asked to depose a plant employee who could answer basic questions: What kind of industrial waste did Washington Works generate? How were these substances handled, at the plant and at its landfills? Where exactly did all those wastes go? Were there any water contamination issues? DuPont's lawyers continued to respond as if I were stepping way out of bounds. These were perfectly ordinary questions. I began to discern a hidden message in their resistance. If I asked for information on anything other than a "listed and regulated" chemical, DuPont would react with outrage. Their lawyers insisted that if I wanted to know about anything other than the materials specifically listed on the landfill permits, I would have to name a particular chemical.

That was a catch-22: I couldn't know the name of a chemical I didn't know. How was I supposed to identify the mystery chemical? I could have had the water tested myself, but analysis is completely useless if you don't know what you're testing for—a lab can't just run a test that tells you everything that's in the water; you have to indicate what specific things you want analyzed. Only DuPont knew what materials they generated at the plant and which ones were going into that landfill. They should be telling me what chemicals were involved—not the other way around. Why were we wasting time with this stupid guessing game?

In the year after I began requesting documents, DuPont had sent me nearly sixty thousand pages of materials it deemed to be within the "proper" scope of discovery. All of them focused narrowly on the permitting of the landfill itself and the chemicals specifically listed in the permits. Not a single page gave a hint of anything in the landfill that could be killing Earl's cattle.

"We'll get to the bottom of it," I had promised Earl—more than a year before.

And still I had no answers.

Time was running out. January 3, 2001—the new scheduled trial date—was now just six months away. I couldn't bring myself to tell Earl: I feared we had no case.

5

THE SECRET INGREDIENT

August 2, 2000
Cincinnati, Ohio

On my thirty-fifth birthday, a mail-room messenger rattled a metal cart into my office. On it was a cardboard box the size of a mini-fridge. My battle to force DuPont to produce records from the plant files relating to *all* chemicals dumped into the Dry Run Landfill had finally borne fruit. We heaved the box off the cart and plunked it down in the middle of my office floor. As the mail-room guy left, I sliced open the heavy tape and peeked inside: twelve hundred more pages of documents, jammed together in no apparent order—a little birthday surprise from DuPont. I poured myself a fresh cup of coffee, shut my office door, and set my phone on "do not disturb." I peeled off my jacket, plopped down onto the floor, and dug in.

It was the nineteenth box from DuPont in the last nine months. Reading and arranging the sixty thousand documents that had preceded these had been one of those boring tasks that other lawyers dreaded and often shunted off on low-level associates but that I secretly enjoyed. It was sort of like detective work, following a paper trail, hunting for clues. They appeared out of order, in disparate fragments that I had to piece together. I'd sketch out the big picture with documents of obvious importance, but that left major gaps I'd gradually fill in over time as I began to grasp the context.

I had my own idiosyncratic system. First I'd spread the contents of a box all over my office floor. Then I would put everything into chronological order. Next I'd read them, tagging documents with color-coded sticky notes according to topic or theme. They would be copied and cross-filed in my windowless storage room by Kathleen Welch, a paralegal who worked for our environmental practice group. Shy, dedicated, and profoundly well organized, Kathleen was a few years older than I and one of those amazing people who could juggle a dozen things at once with speed and something close to perfection. She was slender with curly light brown hair, friendly, but like me not much of a schmoozer. She preferred to keep her head down and focus on work, of which she had a never-ending supply. I alone could have occupied her for a standard workweek, yet there were multiple attorneys who needed her organizational talents. I never once heard her complain about being overworked. Indispensable in cases with lots of files—and my files were beginning to multiply like rabbits on Viagra—she could find anything at a moment's notice. We called her "the document wizard."

Other than Kathleen, I had become pretty much a one-man show. Although Tom and Kim were always happy to talk with me, I didn't have a big team working with me on the Tennants' case to share the research and late-night reading marathons. Some nights it was just me and DuPont in the room. It was wearying work. I often had to battle to keep my eyes open. I would tell myself: finish this file; then you can walk across the hall to the coffee station and have another cup. I must have logged my ten thousand steps every night just journeying back and forth across that hallway. Fortunately, my capacity to absorb caffeine without ill effect seemed unlimited.

The truth is, there was something I preferred about working alone this way, late at night, when there were no distractions, no phones to answer or even anyone to converse with. Some lawyers would have had a roomful of associates or paralegals to dig through the files, but I needed to see it all myself, to understand the full context of the documents, not just an executive summary. As the process absorbed me, I began to feel I was in my element, in exactly the right place at the right time.

I had never dreamed of becoming a lawyer. Growing up, it had never even occurred to me. I loved art and design. As a kid I was always drawing elaborate pictures of cities, obsessed with every detail and the way the streets and buildings fit together. For a long time, I wanted to become an architect. Then I took a mechanical drawing course and learned that being an architect was all about measurements and geometry. Math and I have never gotten along. Today, I don't even reliably remember my multiplication tables. Numbers give me hives. Reluctantly, I concluded that architecture just wasn't going to be my thing.

Then, in college, I discovered urban studies. I was fascinated by the interplay of capital, technology, and bureaucracy and how these things cause a city to grow or stagnate. Cities are complex and multifaceted, like giant social engines in which each gear is a system that must engage with other systems. The design of those systems—individually and collectively—was a kind of logical architecture, a puzzle I found appealing.

I had chosen to attend New College, a small school (in those days, around five hundred students) in Sarasota, Florida. One of the most liberal of liberal arts programs on a subtropical bay-front campus famous for its wild psychedelic scene back in the 1970s, New College allowed me to create a highly specific major and design a custom curriculum. I majored in urban studies and chose classes in urban anthropology, city planning, and the sociology of communities. There were no grades at New College; classes were all pass-fail. But the rigorous academic standards required self-motivation and a lot of independent study. The classes were so small that our professors really got to know us, which was fortunate for me. I might have disappeared in a program filled with bright, outgoing students. My professors saw past my reticence and helped me recognize and develop my strengths: an affinity for complex topics and a capacity for original thought.

After I graduated from New College, I decided to become a city planner. I applied to several schools that offered a master's degree in public administration and was accepted by Syracuse, Columbia, and George

Washington. (I was wait-listed at Harvard, but I couldn't have afforded it anyway.) A couple of schools offered me scholarships, which helped narrow the options.

As I was weighing my choices, my father urged me to consider law. It was the second career he had chosen for himself, after twenty years in the air force. My dad had graduated from the University of Dayton law school the same year I graduated from high school and had eventually become an assistant prosecutor for the City of Dayton, Ohio. "City planning is such a narrow field," he said. "What if you find you don't like it? A law degree will give you more options." He and my mom convinced me to at least apply. When I did even better on the LSAT than the GRE, that sealed it. Several schools accepted me, but I chose Ohio State for the in-state tuition and scholarship and to stay close to my family. It was only about an hour drive between Columbus and Dayton.

In law school I quickly figured out that I wasn't cut out for tax law or contracts. They bored me to death. So much of the legal profession revolves around abstractions, but environmental law was grounded in concrete realities—soil, air, water, waste, things you could see and measure. Though I've never exactly been a tree-hugging environmentalist, I've always loved nature. When I was a small boy, Grammer would take me on walks, teaching me the name of every bird and tree along the way. As an older boy, I freaked out when my parents pruned branches off a tree in our yard. Years later, at New College, among all the long hair and tie-dye of the Reagan-era hippies, I came to be known as "the conservative." I actually fell somewhere in the middle on the political spectrum, earning the nickname more for my squared shoulders, short hair, and polite manner—basically the bearing you would expect of someone raised on military bases. Environmental law seemed like a nice outlet for all that New College ideology.

After passing the bar, I didn't really think much about whether to become a plaintiffs' lawyer or go into defense work. I just knew that the best and brightest students usually got hired by firms that specialized in defending large corporate clients. Plus, corporate defense attorneys made a steady salary, unlike plaintiffs' lawyers, for whom every case was a gamble. I had student loans I needed to pay off. Besides, I didn't have the big personality I associated with plaintiffs' lawyers.

Though I loved working at Taft, I continued to feel insecure in my job. I worried that my lack of family or school connections that would link me to big corporate clients was a handicap. It didn't help that because I'm shy, I'm often overlooked or underestimated. Even the mostly positive feedback on my work performance was tainted by a consistent caveat: I was too quiet; I needed to make more of an effort to let people know who I was. In short, I needed to be more outgoing, talk to more people, work on my "client development" skills.

For me that would be an uphill battle. Although I often went to lunch with my fellow associates, attended the more formal monthly "firm dinners" for attorneys at the Queen City Club, and appeared with Sarah at the annual black-tie firm "prom" dinner/dances, I did not otherwise have much social interaction with my firm colleagues. I dug into the Tennant case with determination; maybe getting good results for the first client I had brought into the firm would prove to my partners—and myself—that I belonged there.

My chronological piles of documents grew into big stacks—oldest records up top, newest on the bottom.

Now, as I sat cross-legged on my office floor, one document in particular caught my eye. It was a letter from DuPont to EPA dated June 23, 2000—just a few weeks earlier. The sender was someone named Gerald Kennedy. This document was from one of the Washington Works plant files, the files DuPont hadn't wanted me to have. What first snagged my attention was Kennedy's title at DuPont: Director, Applied Toxicology and Health. The recipient, Dr. Charles Auer, had an equally attention-grabbing title: Director, EPA Office of Pollution Prevention and Toxics, Chemical Control Division.

If I had a case, it was going to be all about toxicity.

The subject of the letter was a chemical I'd never heard of: ammonium perfluorooctanoate (APFO). I gathered from the context of the letter that EPA wanted to know if DuPont used APFO—and if so, where. Why? I wondered. I zeroed in on this bullet point in the response:

All of the U.S. DuPont operations that use APFO with significant ex-
posure potential are concentrated at one location; Washington Works
in Washington W.V. Therefore most of the industrial hygiene data and
blood serum data presented in this document are from that location.

Interesting: Washington Works was the same plant that was generating the wastes sent to the Dry Run Landfill, and DuPont was worried
enough about the toxicity of this particular chemical at the plant that it
was closely monitoring their employees' blood.

Thanks to my years working for chemical companies, I understood
that in the world of chemical manufacturing, the people with the highest
exposures to industrial chemicals are often the plant workers who handle
them. Because they have higher exposures, they're frequently the first to
show symptoms of health effects. So plant workers are, in a way, the
chemical industry's version of the mining industry's "canary in the coal
mine." Except that they're humans, not birds.

For the first time in a year, I had that rising tingle of excitement in
my spine, the kind I imagine a prospector gets when he sees a glint of
gold sparkle out of the mud in his pan. I plucked the letter off the floor
and pored over it once more. It referenced the fact that blood samples
had been taken from Washington Works employees as recently as March
and April of that year—so recently, in fact, that DuPont was still waiting
for the results from the lab. And that there were existing human blood
serum data going back to 1981.

But what exactly was APFO?

The letter defined it as "a reaction aid in the production of polytetra-
fluoroethylene and tetrafluoroethylene co-polymers." I had absolutely no
clue what any of that meant. No problem, I'd look it up later. But whatever
it was, DuPont plants, predominantly Washington Works, were emitting
tons of the stuff. According to the document, as of 1999, air emissions
from the plant's smokestacks had been around twenty-four thousand
pounds per year, and on-site water emissions had averaged fifty-five
thousand pounds per year. I understood that "on-site water emissions"
meant "into the Ohio River." DuPont noted in the letter that in the form

of industrial sludge or solid waste, APFO was dumped in three local landfills—*including Dry Run Landfill*. Could this chemical be the needle in the haystack I had been searching for?

I set the letter aside on the floor and, feeling my blood rushing in my veins, finished plowing through the rest of the box. When I found no other reference to APFO, I walked down the hall to the environmental library to flip through chemical dictionaries and run down lists of regulated and hazardous substances. There wasn't a single mention of APFO.

Flummoxed, I called an analytical chemistry expert who had helped Kim in previous cases. He had worked inside a major chemical company and was an expert in forensic chemistry—identifying and tracking down exotic chemicals moving through soil and water. If anyone would know about APFO, it would be him.

He had never heard of it, either.

But, he said, it sounded like a substance he had heard of just recently: perfluorooctane sulfonate (PFOS). He'd just read about PFOS in a chemistry journal. The 3M Company had recently announced—two months earlier, in May 2000—that it would cease production of the chemical.

I had never heard of PFOS, either. My chemical expert launched me down a new rabbit hole, hunting for every bit of information I could find about it. I discovered that PFOS had been invented by 3M in 1948. The compound was used as a manufacturing aid in a variety of products, including Scotchgard, one of 3M's most successful—and profitable— product lines. Why would 3M suddenly halt the production of such an important piece of its financial foundation? PFOS didn't appear on the list of regulated substances. I found no regulatory standards from EPA even mentioning the chemical. There had to be some reason for the company to pull it "voluntarily."

An answer, if you want to call it that, came in a 3M press release on the subject. It made no mention of health concerns; instead it said, "We are reallocating sources to accelerate innovation in more sustainable opportunities."

What was 3M hiding behind the thick smear of public relations gobbledygook? Thank you, *New York Times*. According to an article on

May 19, EPA said that 3M's implication that PFOS's toxicity had nothing to do with the decision to stop producing it was false. In a statement, EPA said the company's own tests had shown that the chemical could "pose a risk to human health and the environment," and if 3M hadn't voluntarily withdrawn it, the agency would have taken steps to force it to do so.

But was APFO, the substance I'd discovered was causing concern at Washington Works, related somehow to PFOS?

The more I tried to nail down a possible connection between the chemicals, the more I was stymied by all the technical and scientific jargon. I was familiar with a lot of chemicals through my work, but much of what I was reading seemed nearly impossible for a lay reader to decipher. I sought tutoring from my chemistry expert. He started talking about chemical derivations . . . acids and anions . . . sulfonates versus carboxylates . . . What I mostly got out of it was a quick reminder of all the reasons I had gone to law school rather than med school.

I'm sure my chem tutor was grinding his teeth by the time I managed to grasp what he was telling me: PFOS was a sulfonate. 3M also made a free-acid form of a similar compound called perfluorooctanoic acid (PFOA). In other words, PFOA and PFOS were basically two forms of the same thing.

And the clincher: APFO was yet another version of PFOA. So the chemical that 3M had been compelled to stop making was a kissing cousin to the chemical DuPont was dumping into the air and water by the ton.

According to DuPont's letter to EPA, APFO was a material that DuPont had been using in large quantities at the Washington Works plant as a surfactant.

Technically, a surfactant is something that reduces the surface tension between two substances, but you can think instead, "makes things slippery." Soap is a surfactant. Soap also famously foams in water—so maybe APFO was what was causing the scum Earl was seeing in his creek? I knew it was a leap, but sometimes where there's the faintest whiff of smoke, there's fire. DuPont had admitted in their letter to EPA that wastes from the plant's manufacturing processes were dumped in Dry Run Landfill and the landfill leached into the creek, yet I hadn't seen any

APFO permit limits or regulatory filings in the thousands of other documents I had been reviewing for the past year.

Then it dawned on me: There *were* no permit limits. There *were* no regulatory filings. Now I got why DuPont had been adamant about limiting my discovery to "listed and regulated" materials.

APFO/PFOA was not "listed" or "regulated."

I went back and reread DuPont's letter to EPA. A new word caught my eye: *Teflon*. Buried in the technical jargon was a mention of PFOA's use in the manufacturing of Teflon.

Teflon was one of DuPont's premiere, signature product lines at the time, accounting for roughly $1 billion in annual gross revenue, or $100 million—10 percent—of the company's annual net profit of $1 billion. If 3M was going to lose hundreds of millions of dollars in Scotchgard revenue by discontinuing PFOS, there was even more money at stake for DuPont if EPA found problems with PFOA.

As I dug deeper, I began to understand what had prompted the letter from DuPont to EPA discussing APFO. After 3M had told EPA it planned to stop making PFOS and related materials, EPA had started asking 3M who else used the chemical and any related products. I assumed EPA must have gotten DuPont's name from 3M as being one of its customers on the related APFO product, prompting EPA to ask DuPont if and how they used the stuff. In their response, DuPont was simply confirming their use of the chemical in response to some narrow questions from EPA.

With federal regulators already sniffing around about PFOS, the last thing DuPont needed was anyone giving EPA any reason to have concerns about PFOA. They certainly wouldn't want EPA to know that a landfill containing PFOA was suspected of making hundreds of cows—and maybe some humans—very, very sick. To me, that explained DuPont's pushback against my discovery requests.

I felt a little sick myself, thinking how close I had come to missing this very important—yet subtle—connection. We had gotten nearly to the eve of trial, now less than four months away, without any inkling of a connection among PFOA, Teflon, and the Dry Run Landfill. Even as DuPont was trying to steer me away from unlisted chemicals, they were quietly working behind the scenes to diffuse concern at EPA. It was getting ever harder to dismiss the conclusion that either the Cattle Team had been hoodwinked by DuPont's withholding of key information or it had been complicit in deflecting attention away from the true issue. Its investigation had covered a lot of things, but APFO/PFOA was not one of them.

I'm a pretty level-headed guy. I rarely get upset. But suddenly my blood was hitting a boiling point. I was partly angry at myself for taking so long to get there. But mostly I was outraged by the callous, cynical game it now seemed to me that DuPont had been playing.

I picked up the phone to call the one person I knew would share my anger. Earl answered. I don't think I even let him say hello before I started rattling off my news. I told him I'd discovered a chemical dumped at the landfill that we hadn't known about before that might be a serious problem—one that DuPont and even EPA were concerned about but that the Cattle Team hadn't even addressed. I was met with silence, long enough for me to wonder if something had gone wrong with the connection. But Earl had just been taking it in, enjoying a private moment of validation before practically taking my ear off. "I told you!" he shouted. "I told you from the start!"

He had indeed.

After I ended that call, my adrenaline surging, I dialed Bernie Reilly's number. It would be a short call.

"I get it," I told Bernie. "I know what this is *really* about."

It wasn't about cattle. It wasn't about being sued by some farmer. It was about something much, much bigger.

It was about protecting Teflon.

• • •

April 6, 1938
Jackson Laboratory, Deepwater, New Jersey

Like everyone else, I knew Teflon as the miracle that made my fried eggs slip easily off a pan onto my plate. A little research revealed an interesting origin story. Teflon, like many "miracles of science," had been invented by accident. It was the product of an epic experimental failure—a fluke that had occurred when a chemist was trying to conjure up an entirely different substance. Such failures are the hallmark of synthetic chemistry, and we can thank stymied chemists for such man-made wonders as Super Glue, saccharin (used in artificial sweeteners), and Scotchgard, not to mention the psychedelic LSD, which was stumbled upon in a failed search for a respiratory and circulatory stimulant.

Teflon, though, is one of the most legendary discoveries in chemical history. Before it was ever used on nonstick pans, Teflon was an industrial chemical created in the late 1930s by a twenty-seven-year-old chemist named Roy Plunkett. Two years into his lifelong career at DuPont, Plunkett's assignment was to find a new refrigerant. Existing refrigerants— ammonia, propane, sulfur dioxide, and chloromethane—were toxic, flammable, explosive, or some combination of all three. Malfunctions could be deadly. Fridge manufacturers were clamoring for DuPont to create an alternative from a class of chemicals called chlorofluorocarbons. Eventually the search would result in "refrigerant 12," later named Freon. But the DuPont team was not there yet. One Wednesday morning in 1938, Plunkett released some gas into a tube in an attempt to synthesize a new refrigerant candidate. The gas, in a small steel cylinder, was called tetrafluoroethylene (TFE). When his assistant twisted the valve to release TFE into the reactor, nothing happened. There was no telltale hiss of gas.

They fiddled with the valve. Even wide open, there was no pressure. Had the gas escaped? Was the cylinder empty? Plunkett placed the cylinder on a scale. The weight confirmed that it still contained some material. He poked a wire through the valve, trying to unclog it. Still nothing came out. Annoyed, Plunkett removed the valve completely and turned the cylinder upside down. As he shook it, flakes of fine white powder drifted down like snow.

He thought he had failed and would have to start over. But first he sawed open the cylinder, curious. Its interior walls were coated with a smooth, slippery substance. He jotted his observations down in his laboratory notebook:

A white solid material was obtained which was [presumed] to be a polymerized product.

On some level, Plunkett understood what had happened. The gas molecules in the cylinder had been like a bunch of loose pearls bouncing around in the container. Under conditions lined up by chance, they had spontaneously snapped together to form a long necklace. The steel cylinder itself may have been the catalyst, triggering the chemical reaction.

Though some chemists might have discarded the "failure" and started over from scratch, Plunkett wanted to know what this new stuff was. He conducted a battery of tests to investigate its properties. What he found was highly unusual. The compound seemed chemically inert; it wouldn't react with anything. A soldering iron wouldn't melt or char it. Water would not rot, swell, or dissolve it. Sunlight did not break it down. Mold and fungus wouldn't touch it. It held up to temperatures that liquefied other plastics. It seemed impervious to industrial solvents and highly corrosive chemicals. It was fantastically slippery—with the friction coefficient of ice on ice. He'd never seen anything like it. Neither had anyone else.

Another chemical experiment had gone wrong in all the right ways. Plunkett had invented Teflon.

Plunkett's discovery was a timely one for DuPont because it turned out to play a crucial role in one of the company's most significant initiatives. As the United States entered the Second World War in 1941, chemical factories nationwide redirected their resources to the war effort. The federal government tapped DuPont's brain trust for the top secret Manhattan Project in an all-out effort to build the world's first atomic bomb. For that, they needed the highly radioactive element plutonium. At the behest of the government, DuPont agreed to build a full-scale plutonium plant in Hanford, Washington. The plutonium production process employed highly corrosive chemicals that ate through gaskets and seals. Only Teflon could withstand all of them.

During the war, DuPont's entire Teflon production was earmarked for government use. Most of it went into the Manhattan Project. One-third was slated for other military uses, such as lining liquid-fuel tanks and making explosives using nitric acid, another chemical that destroyed gaskets. Teflon flummoxed radar, so it was used on the nose cones of proximity bombs (another highly classified technological triumph of the war).

By the time DuPont's plutonium exploded over Nagasaki in a bomb called Fat Man, DuPont was already back to planning for its peacetime operations. Those plans involved a new plastics plant in Parkersburg, West Virginia.

With the war over, DuPont began shifting its business model to focus on chemicals, especially synthetic materials, including two world-changing inventions: neoprene (a synthetic rubber) and nylon. DuPont's leaders doubled down on their search for brand-new materials that nothing else could compare to—or compete with. In other words, the next nylon. Teflon seemed like a strong contender.

But Teflon wasn't like other plastics. The same properties that made it so exceptional also made it extremely challenging to work with. Its white-hot melting point made it impossible to mold or extrude like other plastics. Its nonreactive properties precluded many of the chemical reactions and processes that worked with other materials. And its most defining quality—its nonstickiness, the thing that makes it nearly indestructible—made it tricky to bond with surfaces.

Making the substance was difficult as well. Industrial processes were not supersized versions of what had worked in the lab for Plunkett. The product was prone to clumping, and its properties varied significantly from batch to batch. The ingredient that finally made production efficient and reliable was a surfactant—a soap-like material—manufactured by the 3M Company and virtually unknown outside the industry. What made that substance special was that it enabled the mixing of two substances that do not want to mix. It was a lot like the egg yolk that holds oil and vinegar together in the emulsion we know as mayonnaise. Without an emulsifier (the protein and lecithin in the yolk), mayonnaise would be only vinaigrette—a suspension that eventually separates into oil and vinegar. This new compound was the key ingredient that prevented the clumping that gummed up production.

This secret ingredient that made Teflon possible was PFOA.

6

PAPER TRAIL

January 2001
Cincinnati, Ohio

Discovering PFOA changed everything. Now I had something specific to investigate. The discovery documents confirmed that this mysterious compound was used by DuPont at Washington Works and released into Dry Run Landfill. There was still so much I didn't know. How did it get into the environment? How much of it was present and where? Most important, what were the effects of exposure to it?

The more I searched the scientific literature for studies related to PFOA, the more I wound up at dead ends. Public information about PFOA was somewhere between elusive and nonexistent. It was clear why there were no permits. But where were the scientific studies? Where were all the toxicology reports? It appeared that the only existing PFOA research had been conducted by industrial scientists who worked for companies that manufactured or used it. Many industrial studies are never published, so they're not available to the larger scientific community. After I argued that point to the court, the judge moved our already once postponed trial date back another six months, to July 10, 2001, to give me more time for discovery on this mysterious chemical.

PFOA gave me leverage. DuPont had kept me at bay for months, insisting that if I wanted information about anything outside of "listed and regulated" substances, I would have to specify a particular compound. Now that I could name the chemical, DuPont had a legal obligation to send me all internal records or studies related to it. DuPont continued to resist sending me the additional documents. In another round of a now-familiar dance, I sent a response confirming my understanding that DuPont refused to produce all the requested information on PFOA and threatening to file a motion to compel them to do so. A week later came the response to my threat, which was basically "Go ahead, make my day."

We filed our motion to compel, DuPont filed their response to the motion, which was basically "Enough already!" citing the now nearly ninety thousand pages of documents already produced. Our response to the response: They could produce a million pages and it wouldn't matter if they still didn't tell us everything there was to know about PFOA. All of this ate time. Two months after our initial filing, in a hearing before a magistrate judge in West Virginia, our motion to compel was granted.

Finally new documents began to stream in, bringing the total number to well over one hundred thousand pages. But I noticed huge gaps. Many of the new documents referenced other documents—letters, studies, meeting notes—that seemed to be missing from the files. I requested the missing documents, but again DuPont failed to produce all of them. After months of being rebuffed, Larry Winter and I again appeared in a magistrate judge's chambers in Huntington to argue yet another motion to compel. The court granted our request and ordered DuPont to send me essentially all their internal documents on PFOA. Tens of thousands of new pages poured into my office. Soon my carpet was nearly invisible underneath all the papers, except for a little path I kept clear from my door to my desk. The stacks grew knee high, like a little city—maybe I had become a city planner after all. Some days I found myself sitting on the floor, completely walled in by boxes. When callers couldn't get ahold of me, my secretary had to politely explain that I literally couldn't reach the phone.

The documents I received in this round included toxicology studies, water-sampling reports, and internal studies of workers exposed to PFOA.

These new records were infinitely more useful—if only I could understand them. Even though I had dealt with dense chemical literature every day in my job, my work had focused on permit limits and emissions measurements—not the chemistry of the substances themselves. I struggled to figure out what this stuff was, reading and rereading every page, trying to decipher all the hypertechnical details and tweeze out my specific interests. What was this chemical, and what could its effects on cows be? I didn't feel I could simply hand the questions off to a consulting chemist; in the end, this case rested entirely on the science. So I read and read and read.

All of this while I still had to juggle my steady load as a new partner—more than a forty-hour week in itself, including ongoing Superfund work with Tom, regulatory compliance and permitting work with Kim, and helping our business and real estate partners tackle environmental and insurance issues that popped up in their various deals. I had so many balls in the air that I kept a handwritten list on my desk of to-do items, which I was constantly shuffling based on urgency: what had to be done immediately and what could wait a day. I might be researching something for Kim about acceptable limits for various chemical discharges as I was also preparing for a deposition in a Superfund case with Tom. I spent hours in our tiny environmental library and more hours speaking into the handheld dictation machines we still used back then, followed by reviewing my secretary's typescript of the dictation tape, then proofing the final memo. I often had to travel to the potentially responsible parties (PRP) meetings when one of our clients was involved in apportioning costs for cleanup of hazardous waste Superfund sites, which meant collecting and color-coding an avalanche of file folders to take with me on the plane.

My partners were aware that I was spending an ever-increasing amount of time on the Tennant case, but that didn't seem to affect how much work they continued to send my way. How to handle it? I just stayed longer. That was the firm's ethos, which was unforgiving but also one of the reasons it continued to back me in the Tennant case: once we took on work for a client, we did whatever was necessary to pursue the best interest of that client. Days and nights blurred together as I sorted and read, slipping out occasionally for a cup of coffee or dashing across the street for a bag of popcorn or a sandwich before the downtown shops closed for the night. I lost track of the times I crawled into bed a few hours before dawn. It felt like cramming for final exams—every single week.

At this point, I was barely seeing my wife and children awake. Sometimes Sarah would bring our older son, Teddy, into town, and we'd have a "special lunch" together. And we had the weekends. Except I'd usually go into the office on Saturday for "a few hours" that had a habit of stretching into the entire day. I realized that this was a classic trap for a marriage: the work-at-home mom gets stuck with all the grinding labor of housework and child care, while the workaholic dad gets more and more distant and becomes something of a stranger to his own kids. I also realized I was one of the luckiest workaholic dads in history. Staying home with the kids was Sarah's dream job. She truly loved and was fulfilled by it. And having been a law-firm foot soldier herself, she understood the demands of the job, sympathized with the work I was doing, and approved of it. There was plenty of conflict ahead of me, I knew, and I was so grateful that it wouldn't extend to the home front. That said, I know now that Sarah's contribution was even more heroic than I realized at the time. I was so absorbed in my work for years on end that for her it was much like being married to a man who had gone off to war or was out at sea for long periods. I am humbled by and grateful for her selfless sacrifice and grace. Only recently I heard that she had told a friend, referring to this time period, "I could have walked around naked or on fire, and he wouldn't have noticed."

All I could see were the documents spread out on my office floor like puzzle pieces.

One cross-legged day on that floor, I came across a document that wasn't so technical, written for an audience of ordinary people, not scientists. It was a Washington Works "standby" press release from 1989. Documents like this were typically used as crisis-management tools, prepared in advance, heavily vetted by management, and dispatched only if and when the media or the public got stirred up. This one, it turned out, had never seen the light of day. I read it with heightened interest, trying to see through the public relations veneer to figure out what it didn't say. The document answered a lot of my questions—and opened a Pandora's box of new ones.

It was an announcement of DuPont's purchase of a well field that supplied drinking water to Lubeck, an unincorporated area of about five hundred homes five minutes from downtown Parkersburg and a few miles downstream from Washington Works.

According to the unreleased press release, water testing had been taking place since the mid-1970s, supposedly to ensure that the plant's operations were not contaminating the public drinking water. "We believe that the data indicate that the water is safe and reliable," the document stated.

If the water was safe and reliable, why had DuPont had to buy the well field?

A "backgrounder"— a company-approved script of talking points to be referred to if anyone was ever questioned by media or the public— attached to the document hinted at an answer to that question. A very dark answer.

The backgrounder was written as a Q and A, and most of the questions concerned a substance called FC-143 that had turned up in the well field. The Q and A spelled out a few facts about the chemical (another one I had never heard of).

The document called it a fluorinated carboxylic acid, which meant nothing to me. But it also said it was a surfactant—just like PFOA.

Then I got to the punch line: FC-143 was also used in the manufacture of Teflon.

According to the document, FC-143 was a synthetic chemical manufactured by the 3M Company, and Washington Works had been using it since 1951. That meant that as I read this in 2001, the chemical had been in use at the Parkersburg plant for exactly fifty years.

The third question in the Q and A was "Is FC-143 harmful?"

I held my breath as I read the A: "The issue is concentration—how much and when. Animal studies with rats have demonstrated that it is slightly to moderately toxic."

It added that there were indications among rats of liver toxicity and that "human skin irritations, tearing, and respiratory discomfort" had been noted with "overexposures." It didn't define what an overexposure was but claimed, somewhat contradictorily it seemed to me, that "there has been no adverse effect on employee health associated with FC-143 exposure." In what calculus did skin irritation, tearing, and respiratory discomfort not count as adverse health effects?

. . .

So what was this stuff that had seeped into the water supply?

The Q and A didn't give a satisfactory answer. But it did give me something else, something important. It told me that somewhere, there was actual toxicology data. So why hadn't I seen it in my discovery documents? Under federal law, companies generally have a legal obligation to report any evidence of "substantial risk" to human health or the environment. If DuPont or 3M had conducted studies that had found the chemical was *not* a risk, I hadn't seen those studies, either.

I made a note to ask DuPont for both.

In the meantime, the Q and A did make a shocking admission about the allegedly "safe" chemical, FC-143:

> *Why does it accumulate in the blood of humans?*
> *We don't precisely know the mechanism. We do know that it does not readily decompose, react, or break down. . . . It is expelled from the body slowly.*

Many chemicals are found in human blood. We put some of them there ourselves, such as alcohol, nicotine, and pharmaceuticals. Others get in without our knowing it's happening, through environmental exposure. Most of them are metabolized by our bodies, which break them down into smaller parts (the job of our liver and kidneys), and leave our bodies quickly. FC-143 apparently didn't. FC-143 stuck around for a long time inside the human body.

The Q and A also stated that the chemical had gotten into the water supply by leaching from three collection ponds at the Washington Works plant site, which had been replaced with lined tanks in 1988. That might explain why DuPont had bought the well fields—a you-break-it, you-buy-it situation.

What did all that mean for human health? The draft release stated that DuPont's tests had measured FC-143 in Washington Works employees' blood at levels up to 3,300 parts per billion (ppb). "There has been no adverse effect on employee health at these levels," the document said. It also revealed that DuPont planned to remove FC-143 from plant wastewater and reduce its air emissions—at a cost of $3.8 million.

Question 14 in the Q and A echoed my own thoughts exactly: "If the stuff is not harmful, why are you spending money to reduce air and water emissions?"

> *. . . Even though the material has no known ill effects, it is our intent to minimize exposure which could cause concern associated with accumulation in the blood.*

In other words, the company suspected it could cause trouble if it stuck around in people's blood, and they wrote it down in a "standby" Q and A, which they didn't bother to share with the people who had it in their blood.

Not long after I found the 1989 standby press release about FC-143, I came across a strikingly similar document dated almost three years later, in 1991. This one also had a Q and A, and many of the questions were almost identical—only the name of the suspect chemical had changed: "FC-143" had been replaced with "C8." Because the documents were so similar, I surmised that I was looking at two different names for the same chemical. There was, however, one noteworthy and rather alarming addition to the updated Q and A, an additional question, number 20: "Is C8 carcinogenic?"

The answer was not entirely reassuring: "There is no evidence that C8 causes cancer in humans. Tests with laboratory animals demonstrated a slight increase in benign testicular tumors." As I would soon learn, in the study of cancer linkage, *any* tumor formation is extremely bad news. Once again, the document acknowledged that the chemical was a concern at the Washington Works plant but said that "exposure limits for C8 have been established with sufficient safety factors to insure [*sic*] there is no health concern. Monitoring programs show that Washington Works continues to be well below these limits."

Since C8 was not a regulated chemical, those "exposure limits" would have to have been set by DuPont themselves, sometime around 1989 to 1991—the dates of the two documents. The second Q and A acknowledged that "C8 is present in very low amounts in the air, water, and our non-hazardous landfills." One of the landfills operated by Washington Works at that time was Dry Run Landfill.

The "low amounts" were defined as between 1,000 and 3,000 ppb in the water leaching out of the landfill. Was that a small amount? I had no way to judge. Safe exposure limits are almost always debated and vary widely depending on the chemical. Lead, for example, is considered safe by EPA in drinking water below 15 ppb. The safe level of arsenic is 10 ppb—that is, 100 to 300 times less than the amount of C8 in the water leaching from Dry Run Landfill. Such assessments take years for regulators to come up with. Did I trust DuPont's own internal judgment?

After all, this was in people's water. In Lubeck's water. An image of two of my grandmother's best friends, Flo and Burl Phillips, popped into my head. The photo I had seen documenting my day at the Graham farm? Flo and Burl were in it. They had been a routine fixture at all holiday and birthday celebrations in Parkersburg when I was a kid. They'd even come with Grammer for a visit when my family was living in Germany during one of my dad's air force deployments. I was ten, and I'll never forget the unpleasant shock of seeing the plastic sack Burl had to wear on his hip and of learning what a colostomy bag was. Burl would soon be dead. Cancer is cruel, and then it kills you. Flo died of cancer, too.

Both of them had drunk Lubeck water for decades.

My journey down the rabbit hole had started with PFOA, but through more tutoring from my chemistry expert, I began to wrap my head around the fact that all four of my mystery substances were in fact *the same chemical with four different names.* AFPO (ammonium perfluorooctanoate) was the ammonium salt of PFOA (perfluorooctanoic acid). FC-143 was shorthand for "fluorochemical #143," 3M's internal name for its chemical invention. DuPont's name—C8—comes from the eight carbon molecules that make up the backbone of the chemical's structure. So APFO was PFOA was FC-143 was C8. DuPont referred to them interchangeably. A 1980 company memorandum extolling its unique qualities noted that PFOA had already been used for Teflon manufacturing for more than twenty-five years. The memo noted that "other chemicals have been tested but did not match PFOA's properties." So there were four names for one chemical that DuPont claimed was irreplaceable in the manufacturing of Teflon.

Like Teflon, PFOA was a unique chemical. It didn't easily degrade or break down. But as I would soon learn, the qualities that made it singular also made it singularly dangerous. This dangerous chemical was not only in Dry Run Landfill but also apparently in the public water of the surrounding community.

Now that I understood that DuPont had known for at least a decade that their chemicals were getting into the Lubeck water supply, I had a new target for discovery: the Lubeck Public Service District, which provided the town's water. The documents the district produced provided yet another revelation: DuPont had kept their standby Q and A, and their reason for buying the original well field, under wraps for eleven years. For more than a decade, no one had informed the public about the contamination. It was only in October 2000 that a letter from the Public Service District had finally been sent out to customers disclosing the presence of the chemical in their water. What had changed? Why the announcement after all these years? Within weeks of my call to Bernie informing him that I knew about the Teflon connection, DuPont had begun helping the utility draft a letter to consumers. The letter had downplayed the risks, saying there was no evidence that the amount present in the water could cause any harm to those who drank it.

Once again I called Earl to let him know what was going on. "Brace yourself," I said. "This chemical is not just in the creek but in the public water. Not just yours but that of everyone in town."

Earl's sustained anger reached a new peak: How *dare* they? Why haven't the agencies done anything?

"I guess I'm not just a crazy farmer after all," he said. What did all the folks who'd been angry at him for going after DuPont think of his cause now? There was some satisfaction in knowing his instincts had been right. There *was* something wrong with the water, and there *was* something leaking from Dry Run Landfill. By now, the beginning of 2001, his problem had become the whole town's problem.

And it was becoming my problem as well.

7

THE SCIENTIST

January 31, 2001

After so long in the dark, I felt I had broken through DuPont's defenses, at least sufficiently to finally understand enough about the C8/PFOA connection to Earl's dying cows that I believed I could hold my own while questioning DuPont's scientists on the record. I chose Dr. Anthony Playtis, a fifty-four-year-old chemist from Washington Works, for my first heavy science deposition. I fastened on Playtis because I believed DuPont was still holding out on me in document discovery. I wanted someone I could grill under oath about exactly what DuPont knew about water contamination, what records they had generated and failed to turn over to me. His name had been on a lot of the PFOA water-sampling documents that I did have. In fact, one of the sampling sites had been the faucet in his own kitchen. If anyone who worked at the plant would be keenly aware of the threat PFOA in the water presented, he should be.

On the last day of January 2001, I drove to Charleston, West Virginia, where the deposition was scheduled at Larry Winter's old firm. His former Spilman colleagues, representing DuPont, joined me in the conference room, interjecting occasionally to object to the form of a question (one of the only objections appropriate during this type of deposition).

Like almost all my depositions, this one was taped by a professional videographer, and a court reporter typed nearby, transcribing every word that was said. I had invited the Tennants to attend. Earl, Sandy, and Jack couldn't make it, but Jim and Della were sitting in for the family. Throughout the seven-hour deposition—a marathon for sure, but not unusual in the world of depositions—they sat silently behind me, listening as I questioned the witness on their behalf.

A slight man with graying hair and wire-rimmed glasses, Tony Playtis had a PhD in organic chemistry and had been with DuPont for twenty-seven years. In depositions, sometimes you get people who are clearly evasive, confrontational, nervous, or all three. Playtis was not any of those. Matter of fact, plainspoken, even grandfatherly in manner, he pretty much flatlined through the questioning. Before his current position overseeing occupational health and hygiene at Washington Works, Dr. Playtis had worked in polymer research, including ten years involved with Teflon production. In addition to a community water-testing program in the 1980s, Dr. Playtis had been involved in the testing of DuPont employees' blood for PFOA, which he and all the plant guys called C8. Though he wasn't a medical doctor, I was looking to get him on the record about how this stuff was getting into people's blood.

"There are a number of different exposure routes" to PFOA, Playtis told me. "You can be exposed through inhalation. It can be absorbed through your skin to a limited amount, but inhalation is still by far more important. Then of course you could be exposed through ingestion, and that would be the drinking water."

As for drinking water, beginning as early as 1984, Playtis oversaw the testing of water samples gathered inside and outside the plant, from sources including drinking fountains (supplied by several wells on plant grounds) and his own home tap (supplied by the public water utility). Plant employees were sent out with plastic jugs, which they asked to be filled with drinking water from faucets around the nearby communities: Powell's General Store in Washington Bottoms, the Penzoil station in Lubeck, and Mason's Village Market in Little Hocking, Ohio, just on the other side of the bridge, ten minutes from downtown Parkersburg, among others.

I was curious how Playtis, as a chemist, felt about drinking the water out of his own home tap, which had been showing PFOA levels of 2.2 ppb in 1988. But as my questioning went on, it became clear to me that he was the consummate company man. If I asked, he'd probably say he had absolute confidence that it was safe to drink the stuff. He'd also probably add that it improved the taste. I decided that there was no benefit in giving him the opportunity.

The same year Playtis's home tap lit up the test tube, Washington Works removed the three on-site collection ponds suspected of leaking PFOA into the groundwater. Although I had not found any governmental standards or regulations for PFOA, Playtis's name was on a memo I had found in discovery that mentioned the existence of two internal DuPont-created exposure guidelines for the chemical: one for employees and another for the community around the plant. Two different guidelines were created, presuming two different types of exposure. The Acceptable Exposure Level (AEL) was intended to protect workers exposed to PFOA in the air at the plant over a limited duration of time: eight- to twelve-hour shifts. The Community Exposure Guideline (CEG) was intended to protect a population with more vulnerable members (children, the sick, and the elderly), from more continuous exposure to PFOA in drinking water.

These guidelines—particularly the Community Exposure Guideline—were a very important find. They were the first (and only) points of reference I had for the allegedly safe limits of PFOA. First recommended in 1988—by DuPont's own scientists—the guideline for PFOA in drinking water was 0.6 ppb, which DuPont's scientists rounded up to 1 ppb. The higher the level of toxicity, or health hazard, the lower the safety guideline typically needs to be set. In my experience, 1 ppb was a very low safety guideline for a chemical in drinking water—remember, this was one-tenth the limit for arsenic. One ppb was the equivalent of one drop of water in an Olympic-sized swimming pool or one second in thirty-two years. And it was essentially the same as the lowest level that DuPont's lab was even capable of detecting at the time—around 0.6 ppb. If the safety threshold was the same as the lowest amount that could be detected at the time in water, the chemical must be pretty potent stuff. Or else it might be due to a particular chemical property that set PFOA apart from other toxins: biopersistence.

"You mentioned that it was biopersistent," I said. "What do you mean by that?"

"Biopersistent means that once it gets into the human body, it stays there for a long period of time."

It was also bioaccumulative, which meant it built up inside the body faster than the body could excrete it. So a person's total dose was a matter not just of exposure but of cumulative exposure over time. Since the people with the highest exposure to industrial toxins were traditionally the employees who worked with them directly—despite gloves, masks, and other safety equipment—I knew that DuPont should be monitoring their employees' health.

"Have you ever had any employees die working at the Washington Works plant?"

"All the time."

"How do you know whether those deaths are or are not in any way related to any exposure to C8?"

"We attempt to do that through epidemiological studies."

"Any at the Washington Works plant?"

"Yes."

"Any other than the liver study that was done in the 1980s?"

I was referring to a 1981 study of Washington Works employees I'd found. The study linked PFOA exposure to certain changes in liver enzymes—not a disease in itself but an early (and often reversible) indicator that preceded more serious symptoms. It was the earliest DuPont worker study on PFOA exposure I had found among the company's internal records. The study had been conducted after some initial toxicology tests showed liver effects in lab animals. When the same effects show up in both animals and humans, those results would almost certainly prompt further studies. So I wanted to know: Was there more human data?

"Yes."

"What has been done since then?"

"We have a generalized epidemiology surveillance where once every four years, we'll update our statistics on causes of death of employees," Playtis said. "We also look . . . to see if we are suffering from any elevated rate of incidence of any cause of death or any type of cancer."

"These reports—you say they're done every four years?" I asked.

"Right. They're updated every four years."

I tried to maintain my poker face. Playtis had unwittingly revealed something crucial. I had found only one such report in the more than one hundred thousand pages I had extracted from DuPont. Every four years meant that there had to be more cancer death reports missing from the discovery materials I'd received. Now I knew for sure that DuPont still wasn't sending me everything.

I had also confirmed—thanks to an email Playtis had written in 1999—that, as I had suspected, DuPont had never sent out those standby press releases about FC-143, C8, and PFOA. That email would provide evidence of what DuPont had known in the 1980s and 1990s—and didn't share with the public at the time.

. . .

Through all of the extensive discovery process, I had been able to find only a few human studies related to PFOA in addition to DuPont's 1981 liver enzyme study. One was a 1993 study conducted by the University of Minnesota that found an association between PFOA exposure and increased prostate cancer among male workers at a 3M manufacturing plant. A 1996 study also suggested that PFOA had been associated with DNA damage. But DuPont and 3M claimed in their scientific papers that an "association" in these studies just meant a statistical suspicion that there might be some connection. It wasn't proof of any cause-effect relationship. For that, according to their experts, I needed more. Which is why I was so intrigued when Dr. Playtis mentioned the generalized epidemiology surveillance done on Washington Works employees every four years. It seemed to me that those epidemiology reports had the potential to provide stronger proof of a cause-and-effect relationship.

I had also found a number of laboratory animal studies on PFOA, some dating back to the 1960s. Almost all of the studies had been conducted by DuPont and/or 3M. The two companies had apparently been working closely together for decades on PFOA animal testing. One of the studies back in 1978 had looked at the effects of PFOA on monkeys. All of the monkeys given the highest dose had died within one month. Clinical signs of toxicity had been evident even in the monkeys given the lowest dose. Despite those chilling results, in the two decades that followed I could find no follow-up investigations in monkeys. So when I discovered emails between DuPont scientists alluding to a new, more extensive, ongoing PFOA monkey study in the late 1990s, I paid close attention. Monkey studies are extremely expensive, so they're generally reserved for a second round of research, prompted by worrisome results of studies conducted on less expensive animals, such as rats or rabbits. In other words, monkey studies are a pretty big deal.

In one memo on the new monkey study from 1999, Gerry Kennedy, the DuPont toxicologist who had written the first letter I'd found on AFPO from DuPont to EPA, began like this: "Last week a monkey in the low dose [group] showed rapid deterioration and was sacrificed. No outside cause for the death was detected. . . . Recall that mortalities have occurred [in the high dose] group."

I put down the email printout and rubbed my eyes. Then I read it again. The meaning didn't shift: more than a month shy of the end of that new six-month study, the monkeys were already dying. Some of them had died of their own accord, while others had been suffering so badly that they needed to be "sacrificed"—euthanized to prevent unnecessarily prolonged suffering.

The dots were connecting, and the pattern they were forming was grim.

Another email relating to the same study sent seven months later, in October 1999, just four months after we had first filed the Tennants' lawsuit, revealed more. By the end of six months, four of the six monkeys "were in distress" and had damaged livers. One of the higher-dose monkeys had died, as had one of the low-dose monkeys. The analysis concluded that the cause of death was "unclear." But that was followed by a line that seemed to leap off the page. It was like the Bob Dylan lyric

"Every one of them words rang true and glowed like burnin' coal." Here's what it said: "Consensus is that the death was PFOA related."

If PFOA exposure was killing monkeys in toxicology studies, what did that mean for cows—or humans—outside the lab?

From the discovery documents I'd been studying, I learned that DuPont's and 3M's scientists had been concerned about the spread of fluorochemicals, the class of chemicals to which PFOA belonged, for at least a quarter century. The internal alarm bells had first been sounded at 3M and DuPont after the publication of a scientific paper by the American Chemical Society in 1976. The paper had a curious impetus: Donald Taves, a University of Rochester toxicologist, had discovered two different kinds of fluoride in his own blood. Finding *inorganic* fluoride was not surprising—it is routinely put into the public water supply as an aid in preventing cavities. But the other form, *organic* fluoride, is man-made in industrial labs. What was it doing in his blood?

Dr. Taves wanted to find out. So he teamed up with W. S. Guy, a faculty member of the College of Dentistry at the University of Florida. They gathered plasma from blood banks in five cities and tested them for the presence of both kinds of fluoride. They also obtained records of how much fluoride was put into the water of those cities. As expected, the levels of "regular" (inorganic) fluoride in blood correlated with the levels of fluoride in the public water. But the levels of synthetic fluoride (organic) did not. Where was it coming from?

Guy and Taves noticed that the compounds found in the blood had a chemical structure consistent with compounds "widely used commercially for their potent surfactant properties." In their publication of those find-ings, they hypothesized that the source was industrial fluorochemicals manufactured by 3M.

Although not necessarily a paper others paid any attention to, I had found documents in the files produced by DuPont making clear that 3M and DuPont had discussed the findings and had promptly begun the testing of their own workers' blood for the 3M fluorochemicals. By the end of the 1970s, both companies had found the chemicals in the blood of those exposed workers. 3M had known about the Guy and Taves study for twenty years before something—most likely findings that indicated some "substantial risk" to health from PFOS—moved company researchers to replicate the original study and sample blood from blood banks around the United States. What they found was shocking: PFOS was not just in the blood of their workers who handled the stuff, it was in the blood of the general population, people who lived nowhere near their plants. Two years later, they did the same type of blood bank study, sampling for PFOA, and got similar results. Both chemicals were showing up in the blood of the general US population, all across the country.

I could hardly believe what I was reading. I had to reread it a few times to make sure it was really saying what it appeared to be saying. The whole country, the entire US population, was showing exposure to PFOA in their blood? How was that possible? First Earl's cows, then the surrounding community, and now—could this be right?—the entire country? The room wobbled. This was more the stuff of Hollywood thrillers than real life. We're talking about a toxic, man-made chemical that was essentially in the blood of the entire population with everyone seemingly unaware of it. Was I missing something? Where were the headlines? Why weren't the regulatory agencies scrambling? I saw no signs of concern in EPA documents. Even in the media, which usually picks up health-scare stories, there was only silence. But two national blood-bank studies couldn't be flukes (they have been confirmed since).

I had been so focused on the impact of the chemical on one West Virginia farm and its neighbors that the sudden shift in perspective not only to people near Washington Works but to all Americans—including Sarah and my boys—hit me with shock force. My first thought, as a citizen and a human being: This wasn't just about communities in West Virginia and Ohio that were near the plant. This was about everyone, everywhere. What had been happening at Washington Works and Dry Run Landfill must be happening in hundreds of other places around the world where the chemicals were used and released. Then, thinking like a lawyer, it struck me that the potential liability facing 3M and DuPont if the public—and EPA—really started piecing all this together was almost incalculable. What had seemed to me bizarre and uncharacteristic out-of-bounds behavior by a historically good corporate citizen made more sense if it had understood the PFOA issue as a threat to its corporate existence. The Guy and Taves study underscored that DuPont had known, or should have known, of the potential harm to which they were exposing millions of people for at least two decades. It was unforgiveable.

8

THE LETTER

After nearly two and a half years of work, I believed I had finally figured out what had happened to Earl's cows and could prove it in court. What had changed everything was having won the discovery wars with DuPont. The document boxes had yielded one stunning piece of the puzzle after another, and the story they had revealed was a damning one. No wonder DuPont had gone to such unusual lengths to keep me at arm's length. But now I held the cards, and it was time to make sure they knew it.

In February 2001, at DuPont's request, the judge pushed back the trial date once again, this time to October 2. I doubted that day would ever arrive. Fictional legal dramas almost always build to their dramatic conclusions at a trial, but in real life that's rarely the case. Instead, battles are most often resolved not in a courtroom but in the run-up to trial. Both sides want to avoid a trial if possible; victory is not winning your day in court but getting the other side to give you exactly what you want in a settlement without going through the time and expense of a trial. But forcing your opponent's hand takes leverage, which I finally felt I had. It was time to show the other side that the farmer's lawyer had figured it out and was prepared to explain all the details of the decades-long PFOA story in a court of law.

In normal circumstances, a standard mediation brief would have been sufficient. These briefs are shows of force, meant to persuade the other side to move toward a settlement by outlining all the evidence you'll bring if the two sides proceed to trial. But these were not normal circumstances. There was more at stake than simply getting the other side to accept responsibility for what had happened to Earl and his family. I needed to achieve justice for the Tennants, but I also needed to do something about the PFOA exposure in the public water in the surrounding communities. Winning Earl's case would do nothing to stop the broader PFOA threat to the public, because Earl's farm used its own well for drinking water, not the public water supply. It wasn't only Earl's cows but entire communities immediately downstream of Washington Works full of unsuspecting people who were at risk, too. Even while I had been focused on connecting the dots in Earl's case these past months, I'd also been brooding over how to devise a strategy that could achieve not only a win for the Tennants but set other gears into motion to help the impacted communities, so that DuPont could not simply sweep the matter under the rug if and when they settled their case with Earl.

I decided on a strategy: I would write a mediation brief of sorts—laying out my evidence—but I wouldn't do it in the conventional way. Instead, I would spell it out in a letter addressed to the regulatory agencies—state and federal—with a copy earmarked for DuPont. I had multiple objectives in doing so. It would not only show DuPont the storm that was coming and move them toward accepting responsibility for the harm suffered by the Tennants but also simultaneously alert government agencies about PFOA. I'd make the regulators' job easy. I'd present them the facts and point them to the various laws and regulations that would empower them to take action. Since I had already done the heavy lifting—sorting through and making sense of more than one hundred thousand pages of highly technical documents—I'd save them the trouble by sending them the most important 1 percent.

. . .

As the Tennants' lawyer, it was time for me to show DuPont that they needed to give justice to my clients or they would face a trial that could lead to far worse repercussions. Realizing that I was the only person not working for or affiliated somehow with DuPont who knew about PFOA and the health threat it posed to the communities around Washington Works, I felt the responsibility as a citizen to sound an alarm. Knowing what I knew, I could not sit idly by. I didn't realize it at the time, but sending this letter would mark my first time acting not only as a lawyer on a case but also as a concerned citizen who felt a civic duty to try to warn the public about PFOA. This single step would change the trajectory of my life.

I had amassed so much information about DuPont's decades-long PFOA story that I would have to lay it all out in a letter so that the implications would be emerald clear. That would be no simple task, and it would take me months to whittle down those hundreds of thousands of pages of documents to the essentials. My colleagues would later call it "Rob's famous letter," but "letter" was a bit of an understatement: it weighed twelve pounds—19 pages of text and 950 more pages of documentation.

What kept me going was knowing that if DuPont refused to accept responsibility for the Tennants' damages and we went to trial, the exercise would be invaluable for preparing and organizing my case. I started setting out everything I had learned from digging through the discovery documents, winding the clock back even further to lay out for EPA where Earl's troubles had begun fifteen years earlier.

. . .

Sometime around 1984, DuPont's water testing confirmed that PFOA— a chemical that their scientists had concluded by 1978 could kill test monkeys—had leached into the aquifer beneath the Washington Works plant, which supplied the plant's drinking water. DuPont's scientists speculated that the PFOA had gotten there from three unlined digestion ponds located on the plant property where DuPont had disposed of thousands of tons of PFOA-soaked industrial sludge waste. The sludge in the ponds was showing PFOA levels as high as 610,000 ppb. As if that weren't bad enough, the leachate was migrating through the groundwater to the drinking water wells of the town of Lubeck, which were located, at the time, immediately next to the plant's border along the Ohio River.

After the 1989 sale of its contaminated well field to DuPont, the Lubeck Public Service District began using new wells located about two miles farther downriver from Washington Works. Apparently, nobody in Lubeck knew the real reason DuPont had been motivated to buy the well field. Conveniently for DuPont, some years earlier, the Public Service District had made it known that it was seeking an expanded well field. When DuPont discovered the contamination of the original well field immediately adjacent to the plant, they offered to buy the existing property, citing the years-old request as a basis, and giving Lubeck the liquidity to purchase the larger well field it desired farther downstream. Lubeck got its expansion, and DuPont could now proclaim to any regulators or members of the public who might ever ask that there was no PFOA problem "off site," because the old Lubeck wells were now part of the plant property. After the sale, DuPont instructed employees to stop collecting water samples around town for testing and to destroy unanalyzed samples from the old Lubeck wells. DuPont would later insist that they weren't trying to destroy evidence but simply informing their people that they no longer needed to pay fees to store the old samples.

In 1988, the same year DuPont's own scientists first recommended no more than 0.6 ppb PFOA in water as a CEG (rounded up to 1 ppb), DuPont attempted to remove what they thought was the source of the PFOA contamination by digging up 7,100 tons of sludge from the three pits at the plant and trucking it to Dry Run Landfill, about six miles away from the plant. The company had received a state permit to dump the sludge into the unlined landfill designated for nonhazardous waste, the landfill that drained into Earl's creek. Why was DuPont able to dump a dangerous chemical into a nonhazardous landfill? Because PFOA was not listed or regulated as a hazardous substance, which was at least in part because DuPont had not provided regulators with all the data they possessed revealing the hazardous nature of the chemical. That omission allowed DuPont to move the toxic sludge out of the plant pits and onto the land they had bought from Earl's family.

Soon after dumping the sludge, DuPont learned through their own water sampling that PFOA was leaching from Dry Run Landfill into Dry Run Creek in wastewater with levels as high as 1,600 ppb. That was more than a thousand times higher than the 1 ppb CEG DuPont's scientists had recommended for human drinking water. For five years, DuPont did nothing about it, nor did they tell anyone outside the company. By the summer of 1993, state inspectors were noticing excessive sediment and discoloration building up in the man-made collection ponds that accumulated the ooze from Dry Run Landfill. Shortly thereafter, DuPont opened the drains on the collection ponds so the leachate could flow out of the ponds and go directly into Dry Run Creek—for more than two weeks. The DuPont records did not reveal *why* they did this, but I could only assume that it was to alleviate the measurable and visible problems at the sampling point before EPA inspectors showed up for their next visit. DuPont waited until four months after the draining of the ponds into the creek to respond to EPA's request for further samples. The samples eventually sent to EPA showed little cause for concern. For the moment, at least, the problem had been sent downstream.

By then, Earl Tennant's cows were beginning to die along Dry Run Creek. On top of that, DuPont's plan to make the Lubeck water problem go away by purchasing the old well field and moving the public water wells farther downriver was backfiring. Shortly after DuPont's scientists had finally approved adoption of the 1 ppb CEG for PFOA in water, first recommended back in 1988, DuPont's water-testing results came back from the new Lubeck wells. The level of PFOA found in the new wells, two miles farther away from the plant, was now over 2 ppb—*twice* the internal DuPont guideline for human drinking water. DuPont now knew they had a problem with PFOA in both the public water and the creek water that Earl's cows were drinking. But so far, no one else was being told, and no one else had figured it out.

. . .

In the fall of 1994, DuPont began routinely dumping a new kind of industrial waste from the Washington Works plant into Dry Run Landfill: PFOA-contaminated biocake—the dregs of liquid sludge, filtered and pressed into a solid "cake." By the following spring, in 1995, discolored, foul-smelling water was being discharged from the landfill's collection ponds into Dry Run Creek, where suds and foam gathered nearly knee high.

At some point—the documents weren't clear on the date—DuPont stopped dumping biocake waste into Dry Run Landfill. In reports to the state, the company claimed to be collecting leachate from the landfill for transport back to Washington Works to be treated and discharged into the Ohio River, though it wasn't clear from the reports what was in the leachate or what that "treatment" would consist of.

In any case, it was too late for Earl's cattle, who continued to die in the dozens.

By that time, Earl was pleading for help from every agency he could think of, including the West Virginia Department of Environmental Protection, EPA, and the West Virginia Division of Natural Resources. Those agencies in turn contacted DuPont to discuss the problem. Despite such complaints, DuPont did nothing to disclose to the Tennants the presence of PFOA in Dry Run Creek or suggest in any way that their cattle should not be drinking the water. Instead, they kept silent about PFOA and took the position with the public and regulatory agencies that all the problems with the creek had been caused by high iron sulfide levels that had been fully addressed and completely resolved.

In the fall of 1996, EPA notified DuPont that it was initiating an inspection of Dry Run Landfill in response to the reports of hundreds of dead cattle and deer in the vicinity of Dry Run Creek. Most of these reports had come from Earl, but other farm families had complained as well. On the day the notice went out, Eli McCoy, West Virginia's highest-ranking environmental regulator, swiftly sent DuPont a document that might help the company deflect EPA. It was a "consent order"—a legal settlement of sorts between a government regulator and someone accused of violating the law. As a defense lawyer, I recognized that savvy legal maneuver, one I had learned myself during eight years of helping large corporations stay out of trouble with federal regulators. A consent order with the state—whose regulators were known and trusted by DuPont—was a preemptive move to make the feds go away. In the pecking order of government regulation, if a state agency is already "on the case," the feds usually back off.

Sure enough, the agreement between DuPont and the state purported to address and resolve all pending regulatory problems at the landfill—in exchange for a $200,000 penalty (paid to the state) and a number of re-medial measures intended to upgrade the landfill operations.

The consent order was inked in October 1996, right around the time when Earl Tennant was complaining noisily to regulators (including the feds). Shortly thereafter, Eli McCoy left his state job for a lucrative consulting-firm job. Among his new clients? DuPont. The company had hired his consulting firm to navigate the very agreement he had helped create. Larry Winter had warned me about West Virginia's "revolving door" between government and the industry.

Thanks to the persistent complaints of a certain cattle farmer, the consent order was unsuccessful in diverting EPA's attention. None of the remedial measures appeared to be saving Earl's cattle. This was the period when Earl conducted his autopsies and shot his videotapes. He was watching, helpless and increasingly angry, as his animals suffered miserable deaths. I thought back to the skepticism I had felt when he came into our offices laden with cardboard boxes. Now that I knew the real story—the truth of what Earl had been fighting, the lack of coop-eration or assistance from people we are all supposed to trust—I felt ashamed.

It was only when it became clear after a couple of years that Earl wasn't going to pipe down that EPA finally stepped in. In the fall of 1997, it dispatched a research team to the Tennant farm to conduct a full-blown investigation of Dry Run Landfill and the surrounding areas. It enlisted the US Fish and Wildlife Service to examine the health of the deer and other wildlife. Without much in the way of explanation to Earl, the researchers went on a gathering spree, scooping up soil and sediment samples and filling vials of water from the streams and wells. They clipped grass and plants with decontaminated knives. In the creek, they netted crustaceans, insects, and other invertebrates. They used a backpack-sized electroshocker to stun and gather fantail darters and other fish from the creek. They set traps baited with oatmeal and peanut butter to collect meadow voles, short-tailed shrews, white footed mice, and meadow jumping mice. They plucked earthworms from the dirt. Some twenty-seven taxonomic groups were collected from five sampling locations for testing. The researchers conducted autopsies to look for abnormalities and tested tissue samples for the presence of chemicals from arsenic to zinc.

. . .

EPA's investigation found adverse impacts clearly evident among numerous animals, plants, and other wildlife in the vicinity of Dry Run Creek. But the team was unable to identify any known, regulated chemical as the clear cause of the problems.

Testing for chemicals is not as simple as pouring a sample into a fancy machine and getting a printout of all the compounds present. Chemists typically run analyses looking for certain preidentified compounds, following published, approved analytical methods. Such methods exist for only a small number of existing chemicals. When using sophisticated equipment, such as a mass spectrometer, the process generates a chart that looks like an earthquake seismograph. Each chemical has a unique signature of peaks and valleys. By comparing the pattern of peaks with the seismic signatures of known chemicals, an analytical chemist can narrow down what chemicals are present. But there are many chemicals for which seismic signatures aren't known or readily available, making full identification of the chemicals in a given sample far more difficult. There's even a name for it when seismic signatures appear that the experts don't recognize or can't with certainty identify; they refer to unfamiliar readings as TICs (pronounced like "ticks"), for "tentatively identified compounds." It seemed an odd name for them because the compounds were not identified at all—tentatively or otherwise.

In this case, EPA's analysis generated certain "peaks," indicating the presence of TICs.

The findings of the Dry Run ecological investigation were presented in a massive report, issued in draft form, in late 1997. The report confirmed the deaths of numerous species and the odorous discharge from the landfill and concluded, "The results of the risk assessment support [Earl Tennant's] assertion that effluent from the Dry Run Creek Landfill may be having adverse effects on the ecological communities . . . present on the [farm]." Earl's suspicions had been confirmed, but unfortunately it ended there.

With DuPont still keeping investigators in the dark about all the chemicals they were dumping into the landfill, the team's best guess was that the cattle and deer might have been poisoned by "enriched levels of metals, fluoride, and trichlorofluoromethane that appear to be resultant of the landfill drainage." Most intriguing to me, the report noted that the symptoms observed in Earl's cows were characteristic of fluoride toxicity and that there were "numerous" other compounds present in the creek (the TICs) that could not be identified. The report concluded that any one of those compounds could be hazardous and called for further investigation. Part of that proposed additional investigation would be to seek further information from DuPont.

None of this pointed directly to PFOA, but the reference to mystery TICs and the speculative mention of fluoride—an inorganic form of the fluoride that was in PFOA—must have sent shivers up DuPont executives' spines.

With EPA preparing to dig deeper so it could finalize its report, DuPont took quick action. They proposed a helpful new approach: the Cattle Team. If all the upset had been started by complaints about cows, they argued, why not just focus on looking at the problems in the cows? DuPont likely understood the seductiveness of this idea. The chemical analysis was extremely expensive, and EPA was paying for all of it. A more pragmatic approach—studying the cattle directly instead of doing a broader (and more expensive) chemical and ecological analysis of Dry Run, with DuPont picking up a nice slice of the tab—would have tremendous appeal for the federal agency. It worked. The draft report was shelved. Nobody informed Earl that some experts had worried about fluoride contamination or mysterious, unidentified chemicals in the water. All he knew was that the government folks continued to do nothing to help him. His cows were still getting sick and dying. That was when he picked up the phone and dialed my number.

The Cattle Team began its preliminary work in 1998 without Earl even knowing it existed.

Of the six veterinarians appointed to the Cattle Team (three selected by DuPont and three by EPA), one of them, Greg Sykes, was a DuPont employee who had been directly involved in DuPont's internal studies of the effects of PFOA on animals for many years, including cancer studies finding tumors in exposed animals. Though it was clear that DuPont knew PFOA was in the creek water way above the guideline set by their own scientists, the cattle report that Sykes helped author made no mention of it.

By the time the Cattle Team report finally came out in December 1999—blaming the problems on Earl's animal husbandry—EPA had moved on to other things. The Cattle Team had succeeded in doing what the consent order had not done: deflect the feds. EPA never finalized the 1997 ecological report that had concluded that there was a toxicity problem in Dry Run Landfill. It never returned to do any further study of the area. It never did anything to improve the situation on Earl's farm.

And it most certainly did not get around to identifying those mystery TICs.

In the summer of 1999, when Larry and I were shown around the farm, according to DuPont's own documents, Dry Run Creek was showing PFOA levels as high as 87 ppb—87 times higher than DuPont's CEG for humans.

And that, at the beginning of March 2001, was the broad arc of what

I'd learned so far about PFOA and the hazards it posed for those with the misfortune to have it in their drinking water. I assembled the package of documents and wrote a cover letter to state and federal regulators outlining all the evidence and asking the regulators to step in and address the public health threat posed by the PFOA releases at the landfill and the associated drinking water contamination, including taking "those steps necessary to begin regulating C-8 releases into the environment." I cited a variety of federal and state laws that I believed provided the agencies with the power to fix the problem, including authority under federal law for EPA to order DuPont to "immediately cease all manufacturing activities involving PFOA until DuPont can prove through appropriate scientific research that its usage of PFOA does not pose an unreasonable risk of injury to health and the environment."

Then I flipped over my ace in the hole: "This letter also constitutes notice on behalf of the Tennants and a class of other individuals similarly situated of their intent to bring citizen suit claims against DuPont."

This was an important strategic move. The citizens' suit laws allow private citizens to go after polluters for offenses usually enforceable only by regulatory agencies when those agencies are not enforcing the law. To encourage private citizens to take on such cases, the federal environmental laws authorizing such claims typically allow the citizens to recover all their attorneys' fees from the offending party, if the citizens prevail. Through my letter, I was giving notice that if the PFOA contamination of the Lubeck well fields and the surrounding community was not addressed within ninety days of my letter, I might add such citizens' suit claims to the Tennants' case to be pursued not only by the Tennants but also on behalf of all their neighbors drinking the contaminated water. In other words, not only might I add the citizens' suit claims to the Tennant case but I might also expand the case into a class action seeking relief for the whole community affected by contaminated water. The basis of the new class-wide claims would not be for personal injuries or damages but for DuPont's alleged violation of federal environmental law. If we prevailed on the citizens' suit claims, DuPont would not only have to fix the problem for the whole community but pay all our attorney's fees as well.

I was laying a trap of sorts for DuPont. Based on my years of representing companies in such situations, I knew what DuPont's lawyers would likely suggest the company do in response to my threat: they would seek another consent agreement with state regulators, one that purported to address the contamination in our complaint. Under the citizens' suit laws, a signed consent agreement was often viewed by courts as evidence that the agencies were now "doing something," thereby making it unnecessary for the citizens themselves to enforce the law. I was hoping to push DuPont into the arms of state regulators at the same moment I was opening the government's eyes to a thousand pages of the company's files on PFOA. Armed with all the facts neatly summarized in my letter, I felt sure the regulators would at least force DuPont to do further testing of the water and clean up their mess. DuPont would no longer have sole control over the data on PFOA or the ability to manipulate how it was interpreted.

When I mentioned expanding my suit to include "a class of other individuals," I also wanted DuPont to recognize the repercussions if they continued to refuse to accept responsibility in the Tennant case. They might well conclude that a reasonable settlement offer was the best way to prevent me from using the Tennant case as a launching pad for an expanded class action, which would open them up to far greater liability than a suit by a single farmer. They might also consider a settlement preferable to going to trial, which would make all these documents more readily available to the public and not just to regulators whom they knew and might put on the DuPont payroll in the future.

I mailed my letter on March 6, six months before the new scheduled trial date, sending copies to the heads of several federal and state agencies, including several administrators at EPA and the West Virginia Department of Environmental Protection, the state and federal attorneys general, and the US Department of Justice.

A "courtesy copy" also landed with a twelve-pound *whump* on a desk somewhere at DuPont.

9

THE MEETING

March 26, 2001
US District Court, Charleston, West Virginia

A few weeks after mailing the letter, I was scheduled to speak at a public meeting in Washington, DC. EPA was considering regulation of PFOS—the close chemical cousin of PFOA that 3M had already committed to stop producing. 3M had recently reported some new and very troubling lab results for baby rats exposed to PFOS, which had served to keep EPA's attention focused on the merits of restricting further uses of the chemical by any companies in the United States. Even though 3M had agreed to stop making PFOS, other companies that had been buying PFOS from 3M had stockpiled supplies or could make the chemical themselves, and the meeting was to determine under what conditions and for what purposes PFOS could be used going forward. Part of the process was inviting the public—scientists, industry representatives, citizens of any kind—to voice opinions or make arguments for or against regulation.

This public forum seemed like the perfect time and place to raise my concerns about PFOA. I felt the importance of keeping the heat on DuPont and government regulators in the hope that (1) someone in the government would finally do something to help Earl, and (2) DuPont would finally accept responsibility for the mess on the Tennant farm. But it was also about the new and growing sense of responsibility I felt for waking people up to the alarming set of facts I'd uncovered revealing a more massive, widespread public health threat. I'd grown up as a service brat in a variety of towns and cities, and I'd never thought twice about drinking the tap water anywhere in the United States. It depressed and frightened me to think of all the people whose trust in the simple safety of putting a glass under the kitchen faucet was being betrayed.

But now it wasn't clear that I would be allowed to speak up about PFOA at the conference—or anywhere. Almost immediately after the thud of my twelve-pound letter had sounded, DuPont had volleyed back hard. They had asked for a gag order against me. I was fighting it. The federal court judge in Charleston, West Virginia, who was overseeing the Tennants' case had ordered an emergency hearing in two weeks' time— less than twenty-four hours before I was scheduled to speak at the conference. As I reread DuPont's very aggressive filing, I felt the sting of embarrassment, humiliation, and, most of all, anger. This was a personal attack. Citing passages in my letter, DuPont's attorneys were calling me unethical, saying I was a sleazy, publicity-seeking attorney. When I had received the letter, I'd called Larry Winter and told my bosses Tom and Kim. They said I had been in the right to act as I had. Sarah, as always, was calm, levelheaded, and reassuring. She said, "That's ridiculous. I'm sure the judge will see right through it." But I still couldn't sleep the night the order arrived or many thereafter. I kept replaying my letter in my head, parsing every word over and over.

The fact that I had been notified of the hearing at all was already a partial victory. DuPont had sought an ex parte hearing—under seal. *Ex parte* means a ruling by a judge in a hearing where not all parties are present. It's rare, generally reserved for situations in which someone needs protection from imminent harm. Being under seal meant that the gag order would be hidden from the public record. In other words, DuPont tried to meet with the judge behind my back to secure a secret gag order in a hearing I would not be able to attend. The judge denied the ex parte request, so at least Larry Winter and I were notified and allowed to be present.

On the three-and-a-half-hour drive from Cincinnati to the federal courthouse in Charleston, I seethed the entire way. And worried. If the court ruled in DuPont's favor, not only could I be prohibited from speaking at the meeting and communicating with government regulators but DuPont might also try to use the ruling to disqualify me somehow from Earl's case. So I was nervous, and I had no idea what to expect. Although Judge Joseph Goodwin had been assigned to our case from the start, discovery requests had gone through the judge's magistrate, and we had yet to have much direct interaction with him.

Inside the impressive, newly built neoclassical courthouse, DuPont's three lawyers, all former colleagues of Larry's from Spilman, were lined up against me in the hearing room. I glared at them from across the room as they accused me of breaching professional ethics by attacking them publicly. Judge Goodwin, a man with neatly cut white hair and wire-rimmed glasses, wasted no time.

"I've read what you submitted. I've read the plaintiffs' response. I have a hard time seeing—" he began, then stopped himself just short, it seemed, of telling DuPont's lawyers what he thought of their brief. I tried not to read too much into that.

DuPont's most senior attorney, John Tinney, must have felt the headwinds but proceeded to argue that my letter was a deliberate attempt "to produce adverse and prejudicial reactions." He said I was trying to "poison the well"—an ironic choice of words considering DuPont's ongoing problems with poisoned wells—with facts intended to bias public officials who might be called into court as witnesses. He accused me of violating the Code of Professional Responsibility, which holds that "a lawyer is to refrain from making extrajudicial [out-of-court] statements that go to the 'character or credibility' of a witness."

"What statements are they making that you want to stop in that regard?" the judge asked.

"That DuPont fabricated results; that DuPont engaged in impropriety; that the only reasonable inference that can be drawn from it is they got a consent order through Eli McCoy."

Eli McCoy was the regulator who had left his state job to implement his own consent order in a consultancy paid by DuPont. I had traced that relationship in my twelve-pound letter.

"The inference is that DuPont did something improper, that they bribed somebody," Tinney said.

I had never said anything about bribery. I had simply laid out the facts.

"This chemical"—which DuPont's lawyers tended to refer to as C8—"is a nonregulated chemical," Tinney continued. "An effort to influence and persuade EPA to regulate C8 would be a substantial prejudicial thing to DuPont."

I had to bite my tongue. I was simply asking EPA to do its job.

Tinney continued in somber tones. If DuPont was forced to stop using PFOA in their manufacturing process even between now and the trial, "the impact of that would be . . . catastrophic. The plant would shut down."

I recognized the scare tactic. And I knew it was bogus. I'd seen documents in which DuPont scientists reported having identified alternatives to PFOA that they were already testing.

The judge took all of it in. After a beat, he asked, "Then you would be asking me not to allow a citizen to make a complaint to a federal agency with jurisdiction to determine that?"

"Your Honor, we're not talking about a citizen here. We're talking about a party litigant. And—"

"They're still citizens," the judge said. "I would note that there is a substantial public interest in citizens being allowed to petition their government and to file complaints with appropriate agencies as they see fit," he continued. "Moreover, I'm not persuaded that the defendant in this case has shown any irreparable harm that would arise."

He picked up his gavel and rapped it. "I'll deny the motions of Du-Pont."

"Well done" was all Larry said as he jumped into his car and drove off. Just another day at the front. I wasn't nearly as blasé. I had three and a half hours on the road back to Cincinnati to feel relief but also to wonder just how nasty this was going to get with DuPont before it was over.

. . .

The next morning, I flew to Washington, DC, and drove to the Sheraton (now the Westin), a high-rise hotel about five minutes from the Pentagon. The hotel ballroom slowly filled up with a crowd of around a hundred people, all of them with some vital interest in the future treatment of PFOS, which, like its cousin PFOA, was being used in a dazzling range of products and industries. PFOS could be found in everyday household goods, including cleaning products, firefighting foams, fabric treatments, and metal plating, and was used in aviation and the manufacture of semiconductors. The mix of businesspeople from various industries, government officials, and a handful of lawyers, scientists, and environmentalists reflected PFOS's wide reach. Of the business folks, about a third were from the chemical industry.

Presiding over the meeting was Charlie Auer, EPA's director of the Office of Pollution Prevention and Toxics. Auer was the addressee on the first document I'd discovered in which DuPont had mentioned PFOA. Auer, a chemist, had worked at EPA for twenty-four years. He was in charge of EPA's process of assessing and managing new and existing chemicals under the Toxic Substances Control Act (TSCA, pronounced TOSS-cuh).

Before the passage of TSCA in 1976, new chemicals had essentially been treated as safe until proven otherwise. TSCA had imposed stricter front-end regulation, giving EPA authority to review and assess the safety of new chemicals introduced in the United States. A rigorous premanufacture review was required: manufacturers had to provide EPA with information about a proposed new chemical's toxicity, exposures, and environmental impact.

Only now was I fully appreciating an essential fact I had missed when I was still puzzling over why I wasn't finding reports of toxic chemicals in the Dry Run Landfill. TSCA focused primarily on *new* chemicals. By the time TSCA was passed, there were already tens of thousands of chemicals in commercial use in the United States, including PFOS and PFOA. For all those existing chemicals, a sort of grandfather clause went into effect. For the regulation of those chemicals—such as PFOA and PFOS—everyone was basically left to operate under a kind of honor system. Companies were required to report to EPA any information they obtained that supported the conclusion that a chemical presents a "substantial risk of injury to health or the environment." But in practice, the requirement had no teeth; it was essentially self-policing. There were no specific requirements for how companies tested and evaluated health risks in existing chemicals. If they happened to do so, found a substantial risk, and reported that risk to EPA, that could be a trigger for EPA to put together a proposed rule to restrict or even ban the chemical under TSCA.

In fact, the whole regulatory system in regard to unlisted chemicals, I was realizing, was predicated on the assumption of the willingness of corporations to self-report and self-police. The entire system breaks down unless corporations voluntarily share all relevant data and notify regulators when they discover the potential of a significant health risk—even if doing so might trigger the regulation of chemicals that would cost the same corporations millions or billions of dollars. DuPont's handling of PFOA was just one example of how the law protecting us could fail, but it was a particularly egregious one. I knew from the thousands of pages of documents I'd been reading that despite their own rising internal concerns, DuPont had not provided regulators with all the data they had about the dangers of PFOA. They were providing selected bits of information, but not everything that would have alerted regulators to the actual scope of the problem.

I had always assumed that in the United States, systems were in place to keep us safe from dangerous business practices. As a corporate defense attorney, I saw myself as part of that system, helping to ensure that corporations understood and followed complex regulatory rules. It was deeply disturbing to realize that my baseline assumption might be naive. I felt unmoored as I realized that the problem I'd found with PFOA and Du-Pont might not be a one-off issue but a systemic flaw in the framework that was supposed to protect the communities downstream from Washington Works—and not just those communities but all of us. The bottom line is that the route to full regulation of a potentially hazardous chemical is steeply uphill, so steep that gravity and inertia—not to mention the venality of some corporate executives—may leave all of us unprotected. It is only at the end point of a lengthy, contentious, and highly politicized process that a grandfathered chemical can finally become "regulated" under TSCA. Given this arduous and slow-moving process, it's not surprising that very few "grandfathered" chemicals ever become regulated.

In this case, when EPA began making noises about possibly beginning the regulatory process for the "grandfathered" PFOS chemical, 3M tried to preempt all of that by simply agreeing to yank PFOS off the market. Now, most of the people who had come to DC to speak on the future of the chemical worked somewhere along the remaining PFOS supply chain. Even if 3M was phasing it out, they were there to make a case for why their own companies or industries should be exempt from future restrictions on the use of PFOS. As each speaker stepped up to the podium at the front of the room to make a five-minute speech, I realized the scope of the problem. PFOS was used in different applications across a wide range of industries, and all the companies considered it indispensable. One chemical company executive described it as "mission critical" for makers of aircraft hydraulic fluids, which prevented in-flight fires. A photographic film manufacturer called it "critical" for making and processing photographic and X-ray films. Someone from the Semiconductor Industry Association declared that compounds such as PFOS make a "critical difference in the effectiveness of our latest cutting-edge technologies."

The speakers pounded on the same point. "There are no acceptable alternative chemistries," said a microelectronics company from Silicon Valley. "We have no evidence that there are viable alternatives," said a PhD in chemistry. "We know of no alternate chemistry which will provide adequate protection to the integrity of an aircraft," said a manufacturer of aircraft hydraulic fluids. One specialty chemicals company admitted that chemical alternatives for PFOS *do* exist but cautioned that they introduced slight variations that could cause big problems for industrial users. "Even the smallest change . . . could result in millions of dollars of loss to our customers until they catch up with what went wrong," he said. When his company had heard the news that 3M would discontinue manufacturing PFOS, it had stockpiled a ten-year supply.

It sobered me to see and hear firsthand what I was up against—what anyone who tried to call foul on a chemical considered a modern miracle that made our lives easier and more comfortable was up against. An attack on PFOS was an attack on Our Way of Life, not to mention a monkey wrench tossed into the great US economic engine.

The industry concerns dominated the discussion, until the question period. One man stepped up to the microphone and introduced himself as an airport worker involved in firefighting. "I know firefighters who have used these surface surfactants, have literally bathed in them for many years," he said. "What is going to be the long-term effect [for] all these firefighters?"

Auer looked at him. "Is that a rhetorical question?"

"I hope it's one you can answer."

It wasn't.

"I don't know that we have information on the exposures that result from that," Auer said. "We have information on the toxicology of PFOS in animals, and we do have information on blood levels that have been detected. Beyond that, I really can't answer your question. Sorry."

· · ·

I was scheduled to be the last speaker. Most of the audience, never large, had left by this point. I was speaking to a mostly empty conference room, aiming my comments at a sparse handful of EPA officials sitting up front, but they were precisely the people I'd gone there to address. I noticed that at least one of DuPont's attorneys had stayed around to hear what I had to say. I sketched out the close connections between PFOS and PFOA. I described what was happening to Earl's cows after drinking PFOA and explained that humans were also unknowingly drinking the very same chemical from contaminated well fields. Since EPA was already evaluating PFOS, it ought to be looking at both. "We are asking that the agency expand the scope of the current regulation to include PFOA."

Given the scant and poker-faced audience, I didn't exactly feel as if I'd hit the ball out of the park. Still, I got to deliver in person the same message I'd sent in the fat letter. I also got something else from the meeting, something potentially important: Auer had mentioned that there might be some PFOA testing coming up and that the results would be entered into the public record.

That meant EPA had created a public docket—a repository of all documents related to the issues. I decided it would be a good idea to funnel the nonconfidential information I had uncovered on PFOA into the docket. That way, not only EPA but the general public would have access to all the mounting evidence. As I walked out of the conference room, I reflected with satisfaction that twenty-four hours after DuPont had attempted to silence me, I was getting the important public health threat information to EPA. Surely EPA would have to do something about it now.

. . .

In the wake of the public forum, I couldn't stop thinking about 3M's decision to pull PFOS from the market. There was an important message for me there. Not only had the news shocked the chemical world, it had had major repercussions on Wall Street. 3M had told its shareholders that the discontinuation of PFOS manufacturing would incur a onetime charge of $200 million. The phaseout would also affect about 2 percent of the company's annual sales of $16 billion (that would be $320 million). And the company's workforce would take it on the chin. Around 1,500 3M employees worked in jobs related to perfluorinated products.

But even from a cold business perspective, though the costs of ceasing production of PFOS were steep, there could be even greater costs if 3M *didn't* stop making the chemical. I don't just mean the liability it could face in the form of lawsuits, even class actions. Class action lawsuits could be huge, but even the biggest settlements or punitive-damage awards would be a drop in the bucket when compared to the revenues of a company like 3M. The real threat would materialize if a chemical ever joined the small group of "listed and regulated" substances under select federal environmental laws, which would then automatically render the chemical a "hazardous substance" subject to cleanup under the federal Superfund law. Liability under the Superfund law could be astronomical, if not limitless. The Superfund law held companies liable for the cleanup of hazardous substances, and it had no statute of limitations. And this was strict, retroactive liability—no proof of any "fault" or "intent" or "harm" was required. In the case of a man-made, biopersistent, bioaccumulative chemical being dumped into the environment for half a century, the cleanup and remediation costs could potentially be enough to put even a Fortune 500 company out of business.

That threat of after-the-fact regulation kept many companies highly motivated to test rigorously on the front end, before a chemical went to market. Their efforts often exceeded the minimum legal requirements under TSCA. Giants such as DuPont and 3M had huge internal divisions devoted to toxicology and industrial medicine. They assessed the safety of each product and monitored employees with periodic medical exams to detect any early signs of health effects. Manufacturers like these became the foremost experts on the toxicology of their own chemicals. But the self-policing nature of TSCA also created an inherent conflict, especially concerning preexisting chemicals that had been in the market for years before TSCA. For these "grandfathered" chemicals, companies were in charge of producing—and reporting—data that could later be used against them. Understandably, there was often tension between a company's toxicologists and the business divisions that had to pay the tab for research and then suffer the financial consequences of the data those studies produced. Which made 3M's decision to phase out PFOS stand out all the more. To justify the huge financial setback and the potential for future Superfund liabilities, it must have found out something about PFOS. Something bad.

10

THE COWS COME HOME

April 23, 2001
The Tennant Farm, Parkersburg, West Virginia

Earl had just come home from the hospital when he heard a thundering sound over the farm. Squinting up into a clear blue sky, he saw a helicopter. It was flying so low and close he could read the numbers on the fuselage, but he couldn't see who was inside. So he grabbed his rifle, raised the scope to his eye, and used it like a pair of binoculars. He saw a man with a camera taking pictures of his farm. The pilot, seeing a rifle pointed at his aircraft, quickly banked and thundered away. Della called my office, but I wasn't in. Larry Winter fielded the call from a DuPont attorney, who was red hot. You better calm your client down, she told Larry.

Subsequent calls between lawyers established that DuPont had sent a photographer to take aerial photos of the farm as part of a property inspection related to the pending lawsuit. "It is a federal offense to threaten violence against an aircraft carrying passengers," DuPont's lawyers noted in an email message. "Please be advised that the helicopter pilot has indicated that he will pursue today's incident with federal authorities."

It was just the latest in a parade of insults that had filled the Tennants' lives since the moment they had filed the lawsuit. As part of the property inspection, DuPont's agents had conducted a walk-through of the family's rental properties on the farm, a trailer and two houses they rented to tenants for the income.

The agents had a right to inspect for valuation, but the Tennants thought they had overreached when they rifled through the renters' closets and photographed their personal belongings, angering not only the Tennants themselves but also the lessees. The Tennants had lost a renter because of that.

The Tennants felt harassed not only by DuPont but by (former) friends and neighbors. Ever since news of their lawsuit had become public, the community's animosity had been mounting. Many took the Tennants' battle with the company personally. They were outraged that anyone would sue the corporation that had "done so much for the community."

Earl and Sandy could hardly go into town without having to face the pushback. As they filled their plates at the local buffet, heads would swivel and eyes would glare. In the grocery store, people they knew would disappear down a nearby aisle to avoid them. Or, even worse, people who had once been cordial neighbors would avoid eye contact, making them feel invisible. Sometimes when the Tennants walked into a room, people stood up and walked out. The ostracism even followed them to church. One Sunday, during the devotional, the speaker made an odd announcement: "Certain people in this congregation ought to know that their pastor is a DuPonter." The Tennants found another church, but the reception they got there was no better. They ended up changing churches at least three times.

Even though DuPont was cutting jobs worldwide, Washington Works was still the biggest corporate employer in Wood County. They kept about two thousand people on payroll, which infused the local economy with millions of dollars each year. Everyone in Parkersburg, it seemed, either worked for DuPont or had family who did. The company's long reach touched every corner of the community. They gave grants to the public high schools and donations to the Rotary Club. In the eyes of many residents, DuPont was paternalistic in the most positive sense, the ever-rising tide that lifted all boats. Some people still thought of the company just as they called it: "Mr. DuPont."

. . .

Things only got worse when Jim and Della asked for my help to launch a petition to revoke DuPont's permit renewal for Dry Run Landfill. As they knocked on doors down Gunner's Run Road and every other street they could find within a two-mile radius of the landfill, door after door was slammed in their faces. Not everyone was against them. With persistence and skins as thick as rawhide, they managed to gather around three hundred signatures. Those who refused to sign sneered at them, calling them troublemakers, even traitors.

They became increasingly suspicious of everyone around them. Della often got the feeling she was being watched. At yard sales and during errands around town, she swore she was being trailed by men in suits, always driving unmarked cars. The couple told me that more than once they'd come home to find their personal files and medical records scattered around the room, as if someone had ransacked the place. It was hard for me to know if that was understandable paranoia from the stress of the situation or if someone really was hounding them. But it was real enough to them. After that, they kept the petition and their personal records locked in the trunk of their car.

Meanwhile, Earl was getting sicker. His nose and sinus passages were constantly plugged, forcing him to breathe through his mouth. His ears hurt pretty much all the time, and he itched in some unscratchable spot deep inside his head. During repeated trips to the hospital, the doctors gave him medicine, heavy-duty corticosteroids to help with his breathing, but they did little to ease his discomfort. Most nights he could not sleep, so he passed the long hours at the kitchen table, writing his thoughts in a paper notebook with blue lines.

It is now 2:05 a.m. and I haven't slep a wink. My throat is swollen, my back is swollen, my eyes feel like something is tryi to push them up

through my forehead with a whisker wheeze every time I suck in air in
or out through my nose.

He had filled dozens of notebooks over the years, writing down all his
problems and memories. He expressed himself in block letters, with creative
spelling and imperfect grammar, but the bitterness burned through.

DuPont has distroyed my cattle herd til they are no good but they have
not distroyed my love for the buiteful animals. This is somp'on they can
not take from me. They may have distroyed my good helth but they
haven't stopped me from trying to go. I hope those ass holes that dun
this to the earth, this is what they enharret for eternal life to live in,
sleep and eat from and drink water like they expected my cattle and
other wildlife to eat and drink.

Earl had tried watering his cattle from troughs filled with spring water,
but nothing had improved. We later tested the drinking water in the farm
wells and found that it was contaminated with PFOA, too. The herd
continued to dwindle until he gave up raising cattle altogether. His own
suffering was unrelenting. His eyes ached, and his vision was frequently
out of focus. Every day and night was clouded by the struggle to breathe
through a stopped-up nose and a pounding head. Some mornings he blew
out enough snot to fill a teacup. His Adam's apple felt funny, as though
something was permanently stuck in his throat. His skin had never been
quite right since the day he had driven his ATV across the creek and the
water had sprayed his face and hand. The creek water had burned like
acid. Six months later, a doctor diagnosed spots on his face as cancerous
lesions and had to burn them off.

His GP told him to get out of the Ohio Valley for a while, to see if
it would clear up his congestion. So he and Sandy drove east for a two-
week vacation in the mountains. They pointed their car toward their
favorite destination, the Dolly Sods Wilderness, where the Alleghenies
rippled across the horizon. The air there was cool, stirred by high west

winds that bent the spruce trees east, sculpting them into gnarled asymmetry.

As they did on most trips to Dolly Sods, Earl and Sandy drove up the gravel forest road to take in the views from Spruce Knob, the highest point in West Virginia. Its rounded summit, just shy of a mile high, was crowned with a stone lookout tower. From there it felt as though you could see forever across swaying stands of windswept spruce, interrupted here and there by alpine meadows and boulder fields. But the view and the moment were sullied by the sinuses that kept draining down Earl's throat. He spat out the car window again and again. Would his head ever clear up?

They hiked to Chimney Tops mountain, a shelf of rock jutting out of the hills. It was a place where Earl expected to feel awe and wonder. But his ears filled with mucus, and his hearing came and went. It was hard to enjoy damn near anything.

Two weeks of mountain air did him some good, though. He stopped wheezing. His eyes started to focus a little better, and one afternoon he caught a glimpse of a cougar. But he still didn't feel like his old self again. Before long, the trip was over and he and Sandy found themselves on the lonely road home. As they approached Parkersburg on US 50, Earl noticed that he could smell the air, even through his stopped-up nose. That was good news in a way. He hadn't been able to smell anything for months, much less air.

The bad news was: the air stank.

Of course, the air in Parkersburg often smelled bad, especially when a temperature inversion kept the fumes from the chemical plants from dissipating. They didn't call the area "Chemical Valley" for nothing.

. . .

Earl's daughters, now grown, stayed out of the fray. They stood to inherit all that was at stake, but their father said that this was his battle and he didn't want them involved in the mess. They had their own mortgages to pay and kids to raise.

Crystal, twenty-five, was newly divorced and raising two young children in town. Outspoken and confident, she had fierce hazel eyes that could laugh and blaze at the same time. She loved cattle and longed for the day she could raise a small herd of her own. Until then she worked for a medical center in Parkersburg, taking patients' vitals. The doctor she worked for had DuPonters in the family, so she was careful never to mention the lawsuit at work. It helped that her married last name was Day, so her boss and her colleagues didn't realize—yet—that Crystal was a Tennant.

Amy, a quiet redhead, was twenty-two. She worked as a provider of home health care services to the elderly. She now lived on the farm at the home place, in the house where her father was born. The house had been updated with electricity and running water. It was owned by her uncle Jack, who lived up the hill near the Point, a short walk from Earl and Sandy's. Amy loved everything about living on the farm: the evening music of bullfrogs and crickets, the earthy smell of the creek. Someday, she intended to raise her kids here.

The sisters felt deeply rooted to the landscape that framed their earliest memories. As they got older, they begged to go with their dad and help him with the farm chores.

They watched from up close as their father paid the price of the land in sweat and worry. On a farm there was no such thing as paid time off, no matching 401(k), no lazy weekends. Yet tax season came just the same. There was always something broken to fix, something that couldn't be fixed without enough money to replace it. Success was measured not in profits but in years without a loss. And success depended on so many things that they couldn't control. The difference between a good year and a bad year could be as simple as weather that was either too dry or too wet or if there was mildew in the hay.

Even as adults, Crystal and Amy saw their dad as the toughest man in the world. He jackhammered roads for the highway department and rid their swimming hole of snakes by shooting off their heads with a pistol. He could rope as well as any TV cowboy, and his right hand was a testament to the dangers of that art. One day he lassoed a bull from his ATV, then went to press the throttle with a thumb that was no longer there. It must have gotten caught in a loop and popped right off when the rope drew tight. Earl knew he was not the first or last cowboy to lose a digit this way. He tied the stump with a cord and got himself a helicopter ride to the hospital. The loss didn't stop him from picking berries, but he joked about having to train his nondominant hand to wield the toilet paper.

Pain was just a part of life. The girls remember the day the jackhammer finally threw out Earl's back in 1985. A neighbor drove him home in a Chevy Chevette. When the door swung open, Earl rolled out of the seat and onto all fours in the dirt. He crawled to the house on his hands and knees and told Sandy to call Jack and Jim and get them to feed the cows. After two months of recuperation, he went back to work, where they put him on "light duty," which meant loading hundred-pound bags of calcium onto the tailgate of a truck. His lumbar region never quite recovered. Some days he couldn't feel his legs.

The girls had seen him hurt plenty, but as kids they had rarely ever seen him sick. Until recently, they'd never seen him go to the doctor if he wasn't broken or bleeding. A few years before, his breathing problems had sent him to the hospital. Now he went there frequently, when his lungs couldn't get enough air. As he coughed and snuffled and wheezed, the girls worried.

Despite his health troubles, Earl had not gone away, backed down, or died. He found grim satisfaction in believing that DuPont had underestimated him. He said that the local DuPonters were talking about him with something between disbelief and grudging respect. He wrote about the scuttlebutt he overheard—possibly invented, almost certainly embellished—in his notebook:

> *My nabor told me he was in a conversation with several DuPonters &*
> *the statement was brought up, that dam Tennant must have a better*
> *set of lungs in him [than] any dam horse that ever walked, because if*
> *he didn't we would have dun had him dead by now & we don't know*
> *why he is still living because his lungs should have dun shut down on*
> *him a long time ago.*

By now, though, whether Earl's neighbors loved him or wished him ill for threatening the benefits DuPont had bestowed upon them was increasingly beside the point. All of them were imbibing the same stuff in their water.

. . .

Five months before the scheduled trial date, my strategy to force DuPont into the regulators' arms was beginning to work. I was monitoring Du-Pont's actions as best I could through the Freedom of Information Act, and sure enough, soon after I had sent my letter, DuPont had scheduled meetings with regulators. There was no question in my mind what was on the agenda for those meetings: consent agreements. But in the two months since sending out the letter, I'd heard nothing from the state or EPA directly, and so far I'd seen no action taken regarding the Lubeck contamination. I felt as though I were walking through quicksand. Meanwhile, I still suspected that DuPont was withholding a huge number of documents concerning the water samples they'd tested for PFOA contamination. At least that I could do something about.

From the documents I had seen, I knew that DuPont hydrogeologist Andrew Hartten was the person in charge of the PFOA contamination testing. The best way to find out what water test documents DuPont might be withholding—and if they were intentionally keeping them back, the results must be pretty damning—was to ask Hartten, on the record and under oath, in a deposition.

On the morning of May 2, Hartten, two DuPont lawyers, Larry Winter, and I met at DuPont's corporate headquarters in Wilmington, Delaware, in the legal department's offices, near where Hartten worked. As was the case in many of the depositions of scientists and engineers, my questioning was a long, tedious slog through a swamp of complex technical issues that soaked up hours on either side of a lunch break. I had to sit there and ask the questions, but opposing counsel had an even tougher job—listening to it all and trying to keep their eyes open. Usually they sat in a kind of Zen stupor, with a desultory scratch on a notepad now and then.

Depositions can be a slow burn, at best, with all sorts of questions front-loaded about the deposee's background, job history, and professional qualifications. In this deposition, it wasn't until after lunch that I began getting at what I really wanted.

In the mid-1990s, apparently concerned about potential government monitoring, DuPont began shifting the majority of their PFOA waste from the Dry Run Landfill about twenty miles downriver to the Letart Landfill, which was even closer to the Ohio River than Dry Run and also adjacent to a public well field that supplied drinking water to nearby residents. There were also private water wells nearby. As I suspected, Hartten confirmed, on the record, that DuPont was sampling for PFOA in those private wells and also that PFOA-contaminated effluent from the landfill had leached into the Ohio River.

That was just what I needed. We had already paid for tests on the Tennants' well, and they had proved that their water was contaminated, which was no surprise to the Tennants. Now here would be proof that PFOA dumped in a landfill could contaminate nearby wells, and that DuPont had known that all along.

I was curious about something else I had noticed. Recently DuPont had switched from the water-testing lab they had been using for years to a new one. The results from the new lab didn't match up with the previous ones. Some movement was to be expected, but the differences here didn't make sense. The new test results were consistently higher than the old lab's had been. I wasn't sure what to make of that, but it seemed odd to me.

"Do you know whether Lancaster Labs uses a different . . . methodology to sample for PFOA than was used by [the old lab]?" I asked.

Out of the corner of my eye, I saw DuPont's chief litigation attorney, John Bowman, break out of his somnolence and begin scribbling notes. Although he was an experienced litigator and trained not to react or reveal anything during a deposition that might signal to his opponent that they had hit a nerve, I was also experienced enough to pick up on what I sensed to be a subtle yet important change in demeanor. I seemed to have hit a sensitive spot with my line of questioning. I didn't have to wait long to see just how sensitive that spot was.

As soon as the deposition ended, Bowman motioned to Larry Winter and me to come with him. That was odd, to say the least. Normally, if he had something to communicate to us after a deposition, he would fall into stride with us as we exited the building and talk as we walked. This was clearly different. He ushered us into a room down the hall in DuPont's corporate law offices, where we all took a seat. The conversation was not recorded, so my memory may not be verbatim, but Larry and I interpreted the gist of his message the same way: What would it take for us to stop all this?

Not sure exactly what was being asked, Larry and I simply stated we were not in a position to have that discussion and needed to get to the airport. The whole way there in the taxi, Larry and I wondered what precise type of conversation Bowman had wanted to engage us in.

Less than a week later, I got a phone call from DuPont's legal office asking to talk about a settlement. There was no discussion of lawyers being paid to just go away.

Thanks to an internal memo I dragged out of DuPont much later, I could trace the genesis of the settlement discussions. It seemed that my peppering of Hartten with questions about the water testing had hit a sore nerve. In a memo to his legal team, Bowman wrote, "We know that Billott [*sic*] has requested information from Lubeck water company. If he seeks to exploit this item, he may decide to amend his complaint or file a separate class action."

There were those two dreaded words: "class action." Given the growing threat of the Tennant suit spawning something far worse, why not attempt to stop the bleeding? When you know you are playing a losing hand, you fold; you don't throw more chips onto the pile. In another legal briefing, DuPont attorneys identified two glaring weaknesses that would almost surely sink the company if the Tennant case went to trial. They described them in clipped, just-the-facts language and presented them with bullets:

- C8 in the stream and we never told them.
- Never told Cattle Team and EPA about C8 in the stream.

In other words: DuPont had known that Earl's cows were drinking PFOA all along. They'd kept that information not only from the Cattle Team and EPA but from me and Earl for as long as they could. They'd allowed Earl to keep letting his cows drink the crap—for years—and then kept silent about it.

. . .

Settlement discussions began over the phone. When we thought we were getting close, we agreed to meet in Charleston in person. Turned out we weren't as close as I'd thought. As an attorney, I can't reveal any of the substance of the negotiations, but I can say that it took additional months to get to an acceptable agreement and that the discussions were extremely contentious, sometimes downright angry. I thought the talks had broken down irrevocably more than once. And I understood the emotion, on my side at least. No amount of money would restore Earl's land, his cattle, or his health. But at least DuPont would face the consequences of what they had done and not done. I knew that it was desperately important to Earl that DuPont be forced to confess what they'd done to him, but I also knew that was not the way settlements in these types of cases worked. I wanted the Tennants to consider that even if DuPont admitted no wrongdoing, a settlement would give them some sense of resolution and maybe even validation. But for Earl, that might not feel like enough. To him, the suit was about holding DuPont accountable for releasing chemicals that had poisoned his land, and that included a public admission.

Earl understood that PFOA would be with him for the rest of his life. PFOA testing done in connection with preparing the case showed that PFOA was not only in the Tennants' well but in their blood. All of them had contamination levels three to ten times higher than the "background amount" that 3M's blood bank studies had found in the average American. The whole family. Earl was his usual laconic self when I called with the results. But I suspected that along with his fear for his wife and children must have come some sense of confirmation: nobody could ever again claim it was all in his head.

The family doctor couldn't say exactly what the chemical in his veins would do to him; he had never even heard of the stuff. But it sure as hell couldn't be good. Earl had turned his own cows inside out. He had seen what PFOA could do. And by this point he could barely breathe.

He figured it was probably too late to restore his health, but he wanted everyone else to know the truth—to see what DuPont had done to him.

11

THE SETTLEMENT

July 13, 2001
Cincinnati

Three months before the new October 2 trial date, the Tennants—Earl and Sandy, Della and Jim, and their brother Jack—drove the three and a half hours to Taft. It was a summer Friday, in the low eighties and sunny, but you'd never have known it in the windowless, air-conditioned Gamble Room. There we were, at the same conference table where we had passed around photos of dead cows three years earlier during our first meeting. The same portrait of a long-deceased partner hung on the wall. The Tennants were dressed in church clothes, possibly remembering that I had worn a suit to their farm or maybe just because they recognized the gravity of the occasion. We all knew this could be the culmination of years of work on our part and suffering on the part of the Tennants. There was a settlement offer on the table.

Larry Winter and I presented the Tennants with their options: take the settlement and forgo a jury trial, or turn it down and try to get their case in front of a jury.

Because my legal advice to my clients on settlement options must remain confidential, I can't repeat, or even summarize, any of what was said in that regard. What I can say, though, is this: even in the chilled conference room, the atmosphere was heated. There were wildly different personalities clashing in that room, and they didn't always agree. Earl was focused on bringing DuPont to justice; Jim and Della, who felt most keenly the pressures from the community, were more concerned about getting the whole thing behind them and moving on. Jack, the most detached, was hard to read. He and Jim were more inclined to defer to Earl, though Jim would spend hours on the computer researching one chemical or another. He tended to be silent, but once he got talking it was hard to stop him. Della spoke up often, forcefully and at length, which I could see irritated Earl no end.

I was grateful that Larry was there, an even keel in an emotionally charged room. After laying out the advantages and disadvantages of each option, he and I stepped out of the room and waited in the firm library while the family conferred in private. This was their decision—not ours.

Though the terms of the settlement are strictly confidential, I can explain the issues a lawyer typically instructs a client to consider when deciding whether to settle or go to court. The decision is not as simple as it might seem.

The first thing to consider: What is your claim? What damages have you asserted in your case? If you won your case in court and got everything you asked for, what is the maximum you could recover? Let's say, for example, your claims are for property damage and health problems, both current and in the future. How do you go about assessing those damages?

For the property, you might have a real estate expert assess the market value of the land and anything on it—including livestock, a commodity with its own market value. Say the market value of your land and livestock is $100. That would usually be the *most* you could recover—if and only if the jury concluded your property was *completely* devalued. The jury could decide that it's only partially devalued, so that $100 value can typically go only one way: down. One possible resolution would be for the defendant to purchase the property at the current (nondamaged) market value. But what if the client wants to keep the property and wants it cleaned up instead? What if the cleanup costs exceed the value of the property? As the plaintiff bringing the case, you often have the burden of proving—under some existing law or guideline—that the contamination is dangerous enough to warrant a cleanup. Then you have to be able to provide evidence of what it will take to clean it up and what that's going to cost. If you're dealing with an unregulated chemical for which there are no cleanup standards, that's very hard to do.

Now, about the health claims. Say you have breathing problems. Let's say you can prove in court that the chemical in question is capable of causing your breathing problems and also that it (and not something else) actually did cause your breathing problems. What's the most the court can say you're entitled to receive? What actual out-of-pocket costs, such as medical bills, have you incurred because of your health problem? Sure, the jury might also be able to award compensatory damages for pain and suffering—if your case ever gets that far.

Once you've calculated the highest amount you could possibly win in court for your health and property claims, you must then assess the strength of your case and the probability that you can win it. In a case of toxic exposure, it may often boil down to a basic negligence claim. In such cases, you typically have the burden of proving four elements: (1) The defendant had a duty of care (for example, a duty not to poison your water); (2) the defendant breached that duty by acting in a way a reasonable person in its position wouldn't; (3) you have an injury; (4) your injury was caused by the breach of duty.

If you fail to prove any one of those elements, your claim is thrown out and you end up with nothing. The defendant usually doesn't have to prove anything. All it has to do is throw stones at your case.

The strength of your suit may depend on the current state of the science. And sometimes the deck is stacked against you. Generally, regulation lags behind science, which lags behind technology. In a toxic tort case, a plaintiff is often told that the scientific community must agree that the chemical you're implicating is capable of causing your disease. But what if the scientific community has not studied the chemical thoroughly? How are you going to satisfy your burden of proof? If the science does exist to support your case, what experts can you bring in to testify? What if the top scientific experts who know anything about the chemical at issue happen to work for the defendant? What if the defendant holds the key to the data you need to satisfy your burden of proof?

Even if your evidence is solid and your case is strong, how much is pursuing it going to cost? You'll need to hire expert witnesses—scientists who can testify in court. Those experts don't just have to be great scientists; they must also be savvy communicators who can translate their (often esoteric) science into layman's terms that a jury can understand. If the defendant is a multinational science company that employs legions of the world's best scientists, your experts will have stiff competition from its experts. So you'll have to hire the best, and they'll cost you dearly. Their rates can approach, and sometimes exceed, $1,000 per hour. And you won't be paying those hourly rates just for the time they spend on the stand; you'll log many hours preparing your expert witnesses, having them research and write expert reports, and doing other tasks. You'll also have to pay for their travel, their meals, and other expenses.

But even if you find the right experts (and can afford them), there's no guarantee that they'll ever set foot on the witness stand or utter a single word before a jury. If you're in federal court, the judge is often the gatekeeper who decides whether scientific expert testimony is admissible before a case goes to trial. Your experts (and their opinions) will typically have to pass what's called the *Daubert standard*, an evidentiary standard intended to filter out "junk science" (theories or opinions unproven by science but presented as scientific fact). As gatekeeper, the trial judge must ensure that an expert's testimony is "relevant to the task at hand" and "rests on a reliable foundation." In other words, is the opinion backed by good science? And is it relevant to your claim?

This is all well and good—unless your expert is presenting a relatively new opinion on something that hasn't yet been thoroughly studied. It's extremely difficult to be the first to present a new opinion in the scientific world. The expert has to break through the barrier of what has been "scientifically accepted." Your opponent will try to get the judge to see it as junk science. If that happens, your expert will be thrown out by the court well before the trial ever begins. And so may your claim.

You must also consider the personal, intangible costs of going to trial.

It can be physically taxing and extremely emotionally draining. A trial can soak up your time and energy like a soul-sucking sponge. Even if you're not personally on the stand, you may have to sit in a courtroom for weeks or months, reliving things you don't always want to relive, often listening to people drag your name—or the names of people you care about—through the mud. There are people whose job it is to attack your credibility, to make you look bad by any means. And they are extremely good at it.

Meanwhile, your maximum recovery amount doesn't go up just because you're spending more money to prove it. All those expenses—not only your own but those of your lawyers and expert witnesses (not to mention their fees)—will be deducted from your award (if you win). If you lose and it's a contingency-fee case, your lawyers will eat those costs—but not your time, your emotional pain, your humiliation.

So every day of your trial diminishes your chances of the eventual recovery, even if it happens, being worthwhile. Defendants know this and use it to their advantage. It's in their best interest to drag the process of going to trial out so long that even if you win, you lose.

Defendants have big expenses as well. They're paying their lawyers anywhere from $500 to $1,000 an hour (plus hotel bills, expert fees, and so on). They're crunching the numbers, calculating their worst-case scenario at trial. What will it cost them to try to defeat your claim? If the most they could get stuck for is $100 but it's going to cost $250 to defend the case, that's an easy call. If the most they could get stuck with is $100 but it's going to cost only $50 to go to trial, they may still prefer to settle out of court rather than face the risk of a trial. It ultimately becomes an economic decision. As they say in *The Godfather*, it's just business.

But what if, for you, it's not about the money—you simply want the defendant held accountable? You might think: I'm going to take them to trial and prove to the world that they're wrong. Well, even if you win, that may not happen. You can't force the defendant to make an admission. They can be found liable, and they can pay the award but still continue to deny any fault.

Moreover, let's assume you win the case and the jury gives you an award. That doesn't mean you will actually get paid. You may win a verdict—but the losing party can usually appeal. The appeal process can often take another year or more. During that time, your costs continue to mount, further eroding your award. And then you could still lose the appeal.

For all of these reasons, around 90 percent of cases are settled out of court. Through the initial litigation process—discovery, exchanging expert opinions—both parties get a better idea of how the case might shake out at trial and what the damages might be, particularly when the judge starts to rule on which experts, pieces of evidence, and claims he or she will allow in court.

Conveying this complex and sometimes counterintuitive calculus can be an agonizing process, particularly for clients, who are likely to come into it with expectations shaped by Hollywood movies. Decisions like this aren't made easily, and there are no second chances. Whatever the Tennants decided would affect their family for decades. I didn't want to rush them.

As Larry and I were waiting in the firm library, I thought about Sarah, who was at home, nine months pregnant, and chasing two toddlers. She had already launched into "nesting mode" and was spending her Friday sorting toys and clothes for tomorrow morning's garage sale. I would not be there helping to sort, label, haul. Just as I hadn't been there for playdates with the kids' friends or trips to the aquarium. Somehow Sarah didn't resent me for it; her ongoing support and understanding were a gift that I couldn't possibly have deserved. I was, and am, indescribably grateful, but the last thing I wanted to do was take advantage of my good fortune. So though I would never have let it enter into the advice I gave my clients, I dearly looked forward to the moment the Tennant case would be concluded—whether by settlement or by trial—and I could spend more time with my growing family.

Larry and I popped into the conference room from time to time to check on our clients. For hours, every time we looked in, an intense deliberation and debate continued to rage. We checked once again as the afternoon was edging into evening. This time the atmosphere had shifted—the reddened cheeks and raised voices had ebbed, the storm had beat itself out, and they seemed to be approaching a decision. Larry and I exchanged hopeful glances. They didn't let us down. The five people had finally arrived at unanimity: they would take the settlement.

Jim and Della, faces beaming with a mixture of joy and relief, rushed over to me to grab my arm and shake my hand, thanking me over and over for "all your hard work" as I congratulated them. But the whole time I watched Earl, silent and sullen on the other side of the conference table, out of the corner of my eye. Clearly he hadn't so much joined in the decision as given in to the others. I was so focused on Earl that I barely heard as Jim and Della offered to take me to dinner.

I was about to accept when my BlackBerry rang.

Sarah was in labor.

Act II

THE TOWN

12

CROSSROADS

August 2001

Sometimes life is a little ham-fisted with the symbolism. The fact that I'd gotten the news that Sarah was in labor at the exact moment the long labor of the Tennants' suit had concluded seemed a little too obvious. But it took me a while to reflect on that. I raced to the hospital in time to witness that ineffable miracle, then stayed with Mom and new son until the wee hours, when the nurses, who had been hinting with increasing frankness that I needed to let Sarah get some rest, finally gave up on diplomacy and kicked me out. Bleary-eyed and happy, I went home as the sun was about to rise on a day when I needed to take care of the garage sale prep and pick up the other two boys, who were staying with Sarah's family. We grabbed some doughnuts and stopped by Dairy Queen on the way back to the hospital; Sarah wanted a "Chocolate Rock," vanilla ice cream with chocolate syrup and almonds. Cradling my newborn son, I introduced Teddy and Charlie to their baby brother, Tony. Teddy tried to give him a doughnut.

That was when the meaning of the odd coincidence of timing hit me. With a new baby at home and closure at work coming with convenient simultaneity, it felt as though life was clubbing me on the head. This should be a time to refocus more of my time and energy on family. I sank into the warm embrace of home, feeling happy and relaxed for the first time in what seemed like forever.

Even as I savored the moment, the path ahead wasn't as clear or as simple as I would have liked. There was still the major issue unresolved by settling the Tennant case: the fact that the surrounding community's drinking water was contaminated with PFOA.

Back when the settlement negotiations with DuPont had been dragging on, I would often call Earl to brief him, but he had been so outraged about the wider contamination of the public water that each time I'd called to talk about his settlement, the thing he had wanted to know was how any settlement we reached could make things better for those people, too. I'd found myself having to explain more than once that the community issue was not going to be fixed by whatever DuPont agreed to do for him and his family. Even if we had gone to trial and won, it would have done nothing to address the PFOA in the public water supply. Earl didn't use the public wells, so he had no standing in that case. That legal nicety didn't sit well with him. "Now, that's not right," he would say in his gruff drawl. "I need to talk at you for a bit here, Rob," he'd continue, his voice edged with the indomitable force of a lifetime spent battling the odds on willpower alone. Then he would indeed talk at me, forceful variations on the theme of justice: you make a mess, you clean it up; you hurt people, you make it right. I knew when he got going that way to let him get it out. "Look here," he'd say. "These folks done wrong, and they need to be held to account. So what if it's not my wells? That's a bunch of legal hairsplittin'."

He was right; I couldn't argue with that. All I could do was try to stay calm and repeat the facts: this was the law, and this was the way the hairs were split, whether we liked it or not. We'd go back and forth like this until eventually Earl would pause, as if finally weary of the obstinacy of injustice, and say, "All righty then, I'll get back at you." Then I'd hear a click, and gloom would sweep over me. Even though I couldn't change the facts, I still felt that I was failing him.

So I found myself at a crossroads. I could call this a win, send the Tennants home with their settlement, if not total satisfaction, and move on with defending corporate clients. Life might actually feel normal again. That sounded really appealing.

I'd done my part. I'd sent my twelve-pound letter designed to alert regulators about PFOA and to leave DuPont no choice but to submit to regulatory oversight in order to shut down the possibility of the citizens' suits I'd threatened to file.

It seemed as though my plan was working. Thanks to my continuing Freedom of Information Act requests, I could see that DuPont was still busily setting up meetings with the regulators. No doubt they were thinking that they were better off cooperating with an easily distracted public agency than facing legal action from me. Soon they'd be joined at the hip. But would that be enough to get the problem cleaned up?

I couldn't stop thinking about how sideways things had gone when DuPont had teamed with federal regulators on the Cattle Team in Earl's case. What if this turned into the Cattle Team all over again? Just because the regulators were now armed with far more information than they'd had in Earl's case, that didn't necessarily mean they would take action or that DuPont wasn't capable of leading them astray. And while I waited for agencies to act, tens of thousands of people were being exposed to PFOA, many without their knowledge. Could I really settle back into domestic and professional bliss while I knew that was the reality?

I began weighing seriously whether the right thing to do was to bring some sort of legal action on behalf of the whole community. That would mean expanding my accidental role as a plaintiffs' lawyer dramatically at exactly the moment my firm might be feeling relief that our experiment in plaintiffs' law had finally concluded. Instead of melding back into the fold of our corporate defense firm, I'd be moving radically in the opposite direction.

. . .

I was exhausted, both physically and emotionally. I had been consumed by the Tennants' case for nearly three years. Every lawyer spends late nights and weekends working on big cases. That comes with the job. But I had carried my full load of normal corporate defense work for Taft on top of Earl's case. I couldn't help but think of the thousands of hours I'd spent leaning over towers of documents when I should have been bent over a toy-train table with my boys. I always made sure to be with my family for birthdays, holidays, and special events, but after the kids had gone to bed, more often than not I went back to work.

When I should have been celebrating the arrival of our new baby and looking forward to a new, smoother path at home and work, instead I was weighing a decision that would throw not only me but my family into uncertainty. I dreaded going to bed at night, knowing I would be unable to sleep as the conflicting impulses—to jump off the hamster wheel of this quixotic battle against DuPont or jump back on—played relentlessly on a loop tape in my head. I was the only one outside DuPont who had ever seen all the documents related to PFOA. Yes, I had dumped the most significant 1 percent of them on the regulators, but I doubted that whoever reviewed them—*if* they reviewed them at all—would fully grasp their significance. I wondered if some of the current decision makers at DuPont even understood it.

I was still wrestling with what course to take when a man named Joe Kiger called and told me a story that helped tip the scales for me, making it almost impossible to say no.

Nine months before he made the call to me, in October 2000—still months before the Tennants and DuPont would begin settlement talks in their lawsuit but within weeks of my call to Bernie warning him that I knew about DuPont's PFOA-Teflon issue—Joe was in the backyard of his tree-shaded two-story home in Lubeck, relaxing on the swing in the white-picket-fence-enclosed garden. He was watching his wife, Darlene, water her hostas and daylilies when they heard the mailman pull up. Darlene went out to the mailbox. She walked back to the yard, sorting junk mail and opening bills. When she pulled out the water bill, it was folded together with an official-looking notice.

"Honey, we got a water bill," she told Joe. "It says something about something being in the water."

"Bring it here," Joe said.

It was a one-page letter from the Lubeck Public Service District—a letter DuPont had pretty much dictated. It contained a lot of technical terms that meant little to the fourth-grade PE teacher. Something about parts per billion, safety levels and guidelines, concentrations of chemicals in the water. Joe read the letter a couple of times without really under-standing it. The language was mildly reassuring, though, so he didn't feel alarmed. He tossed the letter onto his desk in the basement and forgot about it when it got buried under other papers.

A few weeks later, Joe and Darlene were having dinner with a friend. The friend mentioned that her granddaughter, five or six years old, was having problems with her teeth. They were turning black, and no one knew why.

A week after that, Joe learned that a friend across town had been diagnosed with testicular cancer. That made him think about a few other young men in the neighborhood, men in their twenties, who had also developed testicular cancer. Wasn't that a rare kind of cancer? Then he learned that his next-door neighbor, a young woman who was also a teacher, was fighting another type of cancer. Come to think of it, cancer had been making its way through the neighborhood dogs. The folks across the street had recently found both their dogs riddled with tumors. Co-incidence? Joe Kiger thought about the letter in the water bill. It had said something about chemicals in their drinking water.

He went down into the basement, dug through the papers on his desk, and found the notice from the water district. This time he read it closely. There were a lot of words and terms he didn't understand, and it made him feel uneasy. It said that the chemical, called C8, was present in his water at "low concentrations" but that "these levels are below the DuPont guideline and DuPont has advised the District that it is confident these levels are safe."

Joe had so many questions. What was the stuff? If it was so safe, why were there exposure guidelines? Why would the water district write him a letter about it? And why on earth was a chemical plant advising a water district?

Joe Kiger picked up the phone, dialed the DuPont plant, and asked to speak with someone who could answer his questions. He was transferred to Dawn Jackson, the public relations spokesperson for Washington Works. They talked for half an hour, and he felt no more reassured. His sense was that he was being handled. She suggested that he contact Gerry Kennedy, a toxicologist at DuPont's corporate headquarters in Wilmington, Delaware. Kennedy seemed more than happy to talk to Joe and good-naturedly assured him that there was no problem with the water. Nothing at all to worry about. They chatted for a good forty-five minutes. Darlene was standing in the door of his office when Joe hung up the phone.

"Well," she said, "did he answer your questions?"

Joe frowned. "I was just fed the biggest line of bullshit I've ever been told in my life."

"What do you mean?"

"Something is wrong," he said. "They're covering something up, and I have to find out what it is."

It ate at him. He thought about it every day. So he started making more phone calls trying to learn something—anything—about C8. He tried the West Virginia Department of Environmental Protection. Once again he was told that there was "nothing to worry about." And that made him worry even more. He called the Department of Health and Human Resources. "They just about cussed me out," he said. "Just for asking." Word must have already gotten around that he was trying to rock the boat, he decided. "DuPont's tentacles run very, very deep," he told me. He realized that a lot of his neighbors worked for DuPont, so he would have to be careful.

Then Joe called EPA's regional office in Philadelphia.

"I have something going on here," he said, "and I'm not exactly sure what it's all about." He sent EPA a copy of the notice from his water bill, the one that said that C8 was in his water but the levels were safe. The EPA man came back and said, "Joe, I don't know what this chemical is doing in your water. It's an unregulated chemical. Let me get back to you on this." By that point Joe was expecting a bureaucratic runaround. He wasn't holding his breath.

Two weeks later, he was surprised when someone from EPA actually called him back. "I'm going to mail you some documents," the man told him. "You should read them carefully."

Then, before he hung up, he said, "You may want to contact a lawyer."

Joe received the package the next day. One of the documents cited my twelve-pound letter and included the corporate letterhead of Taft Stettinius & Hollister LLP, along with my phone number. For the umpteenth time in several months, Joe picked up the phone.

• • •

Neither Joe nor his wife had any reason to believe that their water had made them sick—yet. But they had to live with the knowledge that for years they'd been drinking water that might make them sick in the future, water they were *still* drinking. The Kigers' plight moved me. It put a human face on the abstract question I'd been wrestling with. The ramifications on my own life for taking on such a case receded. It was simple: Joe and his neighbors needed my help, and I wanted to help Joe and his neighbors. After all those restless nights, the answer was clear: I felt a duty to try.

I had no illusions. This would be a very different case from the Tennant case—a huge leap in the level of time suck, financial commitment, and risk. I was mortified by the thought of what I'd be asking my firm to bear. As far as anyone could remember, Taft had never filed a class action. I could be risking my ability to do future work for my corporate clients. The risk was not mine alone; I'd be asking Taft to venture deep into uncharted waters with a case based on exposure to an unregulated chemical and to do so on the "wrong side of the tracks," potentially alienating the firm's corporate client base.

DuPont had battled fiercely against a suit brought by a single farmer and his family. Their defense of a class action case would make that look like a lovers' spat. Instead of one family of plaintiffs, there would be tens of thousands. Since DuPont had kept a tight lid on their limited investigation of the health consequences of exposure to PFOA, little to nothing was known about the chemical beyond internal industry studies, meaning that on top of the steep legal challenges, I'd be responsible for somehow advancing the groundbreaking science, too. Proving that PFOA was a threat to human health would require the extensive assistance of world-class experts—paid by the hour at the firm's expense. And then there was my own time, which really belonged to my firm. I wouldn't be able to work on much of anything outside this case. Whereas the Tennants' case had produced a small mountain of unbillable hours, a class action case would pile up a Matterhorn of them.

None of that fit with my idea of myself. I had never been someone who felt called to defend the weak or battle injustice. I had traveled through life rather myopically focused on making the grade, getting the job, fulfilling expectations. But now I saw that stumbling into Earl's case through dumb chance had changed me. I'd found a connection with Earl, one that may have started as an echo of the connection I'd had with my grandmother and some happy moments in my childhood but had grown beyond that to genuine caring and sympathy for his struggle. My eyes had been opened to a harsh reality that I had never fully credited: the ruthless exercise of economic power at the expense of the powerless.

Now I found in myself a desire to help, a crushing sense of responsibility to people like Joe, but also the inspiration of Earl's unshakable resolve. Even though Earl and his family had nothing more to gain personally, Earl was the loudest champion of the cause. He wanted to see justice done. "They shouldn't get away with this!" he growled whenever I talked to him on the phone, which was fairly frequently as I tried to sort things out.

I felt as Earl did, not only as an attorney and a citizen but as a man whose own family I increasingly believed might have been directly harmed. Grammer's friends Flo and Burl had been drinking PFOA in their water for years until they had died of cancer. My great-grandfather, who had passed away in 1977, had also been drinking the local water for years until he succumbed to prostate cancer. An increased incidence of prostate cancer had been singled out in some of the studies of workers I'd seen as being associated with PFOA exposure.

I was like Earl in another way as well: I was tired of being ignored. I had done everything I could think of to prompt action by federal and state agencies. Thus far they had taken no action to clean the stuff up, stop its release, or protect people from further exposure. I wasn't some bleeding-heart liberal. And I certainly had no hero complex. But I couldn't unlearn what I knew, and I couldn't just walk away.

I kept coming back to the reason we had taken the Tennants' case in the first place: no one else would help them. Now the same was true for the people drinking the toxic crap in their tap water. Every time they put on a pot of coffee, cooked up a stew, told their children to brush their teeth before bed, or sipped from the faucet when they awoke thirsty in the middle of the night, they were enrolling their families as guinea pigs in an uncontrolled experiment in human toxicology. Neither DuPont, the local water district, nor the government was offering them any assistance.

That's where I hit a wall in my hypothetical case. Given the facts, it wasn't at all clear that a civil suit in these circumstances could help them, either. In most instances, class action suits like this are undertaken only when there is clear evidence that members of a community have been exposed to excessive levels of a regulated substance that the scientific community generally agrees is capable of presenting a risk of serious harm. Here we had an unregulated chemical, so there was no official safety limit or governmental standard in place to exceed. Similarly, since scientists outside the industries that worked with PFOA didn't even know about it, there was no shared consensus outside those companies about what kind of harm it could cause. It wasn't simply long odds. To my knowledge, a class action based on potential harm from an unregulated chemical had never been attempted before. Attempting to try this case wouldn't just be exponentially more difficult, it might actually be verging on insanity.

Without any of the facts established, any lawyer, myself included, would wonder how to even mount such a case, much less successfully prosecute it. Attempting to hold DuPont legally accountable for this mess could drag on for years with no guarantee of success, even as the people who had been unwittingly drinking the stuff (for decades in some cases) were continuing to expose themselves. Every day that lawyers, regulators, and scientists duked it out could continue to harm them and their families. People had a right to know the full extent of their exposure to this chemical, even if we weren't yet sure of everything it was doing to them—*especially* because we weren't sure of everything it was doing.

From my conversations with Joe, who had canvassed his neighbors, I was pretty confident that what the community really wanted wasn't money, at least not primarily. Unlike Earl's cattle, most of those people hadn't even suffered any obvious symptoms that they thought might be related to PFOA exposure, but they had to live with the fear that someday they might. At this point, they just wanted the truth about what risks they faced, and clean water. Those weren't conventional aims for a lawsuit, and I wrestled for quite a while with whether legal action could fulfill them. I did what lawyers do and hit my law books, searching for useful precedents.

For days I trudged through a bottomless pit of legal decisions. Then late one night, bleary-eyed, I came upon a recent case from the Supreme Court of Appeals of West Virginia, just a couple of years old (1999), that grabbed my attention. This relatively new opinion appeared like the mirage of an oasis conjured up by a man dying of thirst, so perfectly suited to our situation that it seemed unreal. The decision was an opinion from the highest court in the state recognizing and allowing a completely new common-law tort claim to be pursued in West Virginia courts called "medical monitoring." In essence, this new type of claim would allow people who had been exposed to toxic substances but were currently healthy to get free medical testing to detect the onset of any disease that could be caused by that exposure, as opposed to the simple money damages that were traditionally available to people who had already been diagnosed with a disease or injury caused by the exposure.

The standard of proof laid out in the decision included essentially five elements, which I would need to satisfy in order to bring a case: (1) significant exposure to (2) a proven hazardous substance that (3) significantly increases the risk of developing a serious human disease (4) for which a reasonable medical doctor would recommend diagnostic testing and (5) for which such testing procedures exist. In other words, if you could prove that you'd been exposed to a chemical that significantly increased your risk for developing disease in the future, the ruling meant that you could be awarded the right to get ongoing medical testing for that disease now. The idea was to detect a potential disease caused by your chemical exposure *before* that disease was so advanced that it was too late to treat it.

A sixth element was also necessary to prevail on this new type of claim under West Virginia law. The Supreme Court of Appeals said that the exposure had to be caused by the "tortious"—wrongful—conduct of the defendant. I felt confident that I would be able to prove all six elements with respect to the community's exposure to the PFOA released by Du-Pont. In fact, I could make a strong case just with the documentation I already had in hand from the Tennant case. DuPont's own scientists had recognized PFOA as a potential health risk and set the 1 ppb exposure guideline. DuPont's own documentation from their water testing showed levels in community water well above that guideline. The company's own doctors had worried enough about the development of serious human disease due to PFOA exposure to have conducted medical testing of ex-posed workers since the 1970s. And as to the last point, DuPont had wrongfully released and continued to release PFOA into the community's water despite their knowledge of the risks.

The fact that this unusual and progressive medical-monitoring ruling popped up at all, let alone in a conservative state such as West Virginia—which DuPont had just happened to choose as the site for the Washington Works plant fifty years earlier—was a startling stroke of fortune. But there was a catch. First, the ruling was so recent that I couldn't find any evidence that anyone before me had attempted to use it in a case like this involving an unregulated chemical; I'd be the first. And according to various business and "tort reform" groups, the ruling flew in the face of hundreds of years of US tort law that required some "actual" diagnosed disease or injury before someone could sue. Many in the chemical industry were already taking aim at the ruling, claiming it opened the flood-gates to massive legal liabilities for chemical manufacturers. The West Virginia Chamber of Commerce held urgent meetings to find some "legislative fix." My reading of the situation was that the ruling was de-spised by many and might soon be gutted when it was attacked in the federal courts (and I would later be proved correct).

But for Joe Kiger and his neighbors it was a window of opportunity, one that would soon close.

It was time to act.

. . .

I gathered my notes and presented the case to the firm. Well, I made my case to Tom Terp in his office. Today such a radical proposal would be handled very differently at Taft, with a formal presentation before a management committee. But back then the firm was less than a third of its current size, and things happened more informally. I told Tom what I had in mind. He understood the potential—the certainty—of huge costs, likely several years of my time and the astronomical fees required for multiple expert consultants. He also knew that this time I couldn't go it alone; I'd need a team of lawyers from other firms with more experience in class actions all willing to jump down the rabbit hole with me.

On the other hand, the payoff, if we could get there, could be sub-stantial. The value of medical monitoring for a class of thousands could run into the millions of dollars. And we could seek to compel DuPont to provide clean, PFOA-free water, another immensely expensive undertak-ing. If we were able to pursue the case as a class action and be awarded attorneys' fees based on the value of such "benefits" secured for the class, our fee could potentially be a substantial percentage—typically 25 to 30 percent—of the total, which would be a pretty hefty sum. Another posi-tive: as a defense firm, we had the funds that could sustain us through years of no return. Plaintiffs' firms, which had to front costs for many contingency cases at once, were less likely to be able to invest in a single extremely complex and expensive case.

Though the case was a novel one, Tom recognized that we would start out way ahead of where we had been at the beginning of the Tennant case. Back in 1999, we had known only that Earl's cows were dying and his creek water looked funny. We'd had no idea of the cause and no evidence that DuPont was connected in any way. Now we knew, and had abundant evidence, that PFOA was the problem, that the level of PFOA in the community's water was above the safety guideline that even

DuPont's own scientists recommended, and that DuPont had recognized the exposure many years back and done essentially nothing to report it or protect the people downstream. Of course we both understood that DuPont would move heaven and earth (and hire the best lawyers) to dispute all of this, attack our facts, bog us down in procedure, drag out the process, and make it as costly and painful as possible. (Even so, our initial imaginings would prove unequal to the ferocity and guile of the counterattack that awaited us.)

Tom considered all this carefully, rationally, betraying no sign of tension over the gamble I was proposing. Then calmly, without fanfare, he said yes. With him on my side, my other partners agreed to take the risk.

That demonstrated real trust in me. I was pleased to have gotten the green light, but my stomach hurt.

What had I gotten myself into?

13

FIRST BLOOD

I knew from the start that I would need outside help on this new case. For many elements of the case, there was no precedent. We would need to not only build our legal case but bring to light the science proving that PFOA was in the water and presented a significant risk to community members at the levels to which they had been exposed. This would have been hard enough in any circumstances, but the fact that almost all the existing research was buried deep within DuPont's internal files amped the degree of difficulty to the level of threading a needle in a rowboat during a hurricane.

Our firm had tremendous experience and talent in many things, but handling such an off-the-map class action on behalf of tens of thousands of individuals against a major chemical manufacturer in West Virginia was not one of them. I needed first-rate talent, attorneys who could help me handle a massive, complex case with many moving parts. I needed partners who could work quickly and collaboratively and think outside the normal way of doing things.

First on my list was my partner in the Tennant case, Larry Winter. Not only had he practiced in West Virginia for many years, he came from the same type of big defense firm background and culture that I was used to. We had established a great working relationship during the Tennant case, and that familiarity would be a security blanket as we entered uncharted territory. We still needed to partner with a plaintiffs' firm with experience in handling mass torts and enough resources to be able to share the up-front costs and ongoing expenses. I found the prospect terrifying. It meant that I'd likely be working with that alien tribe, "real plaintiffs' lawyers." I didn't know any. Where would I even begin to look? Fortunately, one of Larry's partners was married to a lawyer whose firm specialized in class action and mass tort litigation. Hill, Peterson, Carper, Bee & Deitzler had handled some of the biggest such cases in West Virginia history, including taking on Big Tobacco. Larry and I presented our case at the firm's office in Charleston, West Virginia. It was one of the state's most successful plaintiffs' firms, and its digs announced the fact clearly; picture a Tuscan villa nestled in the Appalachian hills. The walls were paneled in beautiful dark wood. A stone fireplace dominated the conference room. There was a Jacuzzi on the deck and a fully equipped gym in the basement. And this was all for what at the time was a six-lawyer firm. The message came across loud and clear: if you were good, being a plaintiffs' lawyer had serious potential upside. Being a little lucky didn't hurt, either.

These lavish offices were a far cry from the staid luxury of corporate defense firms such as Taft. As I took in my surroundings, I felt a bit like Dorothy realizing she wasn't in Kansas anymore. But Ed Hill, the partner at the firm who greeted me as I entered, looked like every standard-issue corporate defense lawyer: distinguished, soft-spoken, poised, and impeccably groomed. He was reserved to the point of shyness—almost the negative image of the loud, aggressive, egotistical stereotype I had built in my mind. I thought, "Wow, this is not what I expected." I was surprised and also relieved. Here was someone I could relate to.

By the time we left that day, we'd reached agreement on how fees and expenses would be split between our firms, and Ed had introduced us to his partner Harry Deitzler, a Parkersburg native who would be our even more local contact with the community.

Though I would talk often with Joe Kiger on the phone, I asked Harry and Ed to be responsible for staying in regular touch with him and the others in the community, making sure that potential class members knew what we were doing at every step and that we knew what they were thinking. Those folks could relate to Harry, in particular, as a Parkersburg neighbor. The built-in trust that fact generated proved essential early on, when anyone who supported the class action in the community faced pushback from DuPont loyalists. But it wasn't full-time work for my West Virginia co-counsel, who were also handling other cases. For me it was more than full-time.

I still had my normal workload of Superfund work with Tom and regulatory compliance work with Kim as the new case began. Though this time I would have the benefit of help from my expanded team of West Virginia co-counsel, at the end of the day it was still my case and I had the responsibility to make sure the entire strategy, top to bottom, was sound and executed as perfectly as possible. In the Tennants' case, that had been pretty straightforward, or so it had appeared at first: I'd had to identify what chemical from DuPont's landfill was getting into Earl's creek and killing his cows. In terms of the law, Earl's case had been simple. This was far trickier. This time we already knew the chemical going in, but that meant we also knew it was unregulated, meaning there would be no permit limits or federal or state standards or even guidelines for us to rely on. Proving that such limits, standards, or guidelines had been violated was usually the whole ball of wax for plaintiff lawyers in chemical contamination suits. If no official limits, standards, or guidelines existed, what could we even allege was being violated? And that wasn't the only problem. In traditional claims involving chemical contamination, plaintiffs' lawyers are told that they have to prove that a legally recognized "injury" (such as a diagnosed disease) was suffered that was caused by the hazardous chemicals. In other words, you couldn't sue until *after* you got sick or hurt. My situation was different; I had large numbers of people who had been exposed to a dangerous chemical without their knowledge and plenty of evidence to prove that the exposure had occurred, and at levels above the safety guideline created by the company that had used the chemical. But those people were not necessarily sick (yet). So what law could I argue had been broken?

This was where that medical-monitoring ruling by the Supreme Court of Appeals came in so handy: while we began the long and difficult process of convincingly confirming that not just PFOA in general but the particular amount of PFOA in the local community water could cause physical illness suffered by class members, our initial claim could focus on the *potential* harm of PFOA in the water.

On August 30, 2001, about a month after the Tennant settlement, we filed the class action suit for Joe and his neighbors, focusing on the innovative West Virginia medical-monitoring claim, with Joe being one of thirteen named plaintiffs. Thirteen was more or less an arbitrary number. We needed to have a deep enough bench to make sure that if someone got sick or decided not to participate, the whole case wouldn't go away. Joe and his wife helped direct interested neighbors to our local counsel, who then identified those best suited for the task of representing the whole community as a named plaintiff on the complaint. The chosen had to be willing to participate in discovery, including producing documents from their own files and sitting for a deposition and possibly even going for a medical exam. They also had to be willing to speak for everyone if the class was certified by the court and there were mediation or settlement discussions. Once the named plaintiffs were selected, they nominated Joe to be our main contact for the group.

We intentionally avoided filing the class action complaint in Wood County, the home of Washington Works and the center of DuPont's patriarchy. We opted instead for Kanawha County, West Virginia, where Charleston is located. It was still in the heart of Chemical Valley, but at least (we hoped) we could avoid a trial that would give DuPont a home-field advantage. I'd already seen how DuPonters' loyalty could harden their hearts to—and cloud their judgment about—the suffering caused by poisoned water.

. . .

In September 2001, less than a month after we filed, the World Trade Center towers came crashing down. Anonymous letters filled with anthrax made picking up the mail a frightening enterprise. The entire world was reeling.

At home, my family was shaken by our own personal crises. My wife's grandmother, the matriarch of her family, had died, succumbing to cancer. On the day of her burial, we learned that Sarah's mother also had cancer, a serious type of lymphoma. On top of managing two small boys and a two-month-old infant, Sarah was heroically working through the grief she felt for her grandmother's passing and trying to hold down the fear of losing her mother so she could be fully present to support her during treatment. And she was doing it all while planning to host her entire extended family at our house for Christmas, including nine children under the age of four.

And me? I'd be at work. I somehow managed to get home by 6:30 each night to eat dinner with Sarah and the boys, crawl around on the floor with them, tire them out, and then pore over Richard Scarry books (my childhood favorites—even then I had an eye for detail and unlimited capacity to focus on minutiae). After helping with the bedtime routine, I would head back to the office for what I came to consider the second part of my day. A rooster to my night owl, Sarah was out of gas by the time the sun set. I was just hitting my stride, so back to the office I went, where no phones were ringing, no coworkers were interrupting, and I could concentrate.

The first part of the case would be all about securing evidence that the other side would be hustling to keep out of our reach. I spent countless hours compiling lists of questions, known as interrogatories, and document requests to formally serve on DuPont to kick off the discovery process. These questions were like a fine-meshed net designed to catch all the information—helpful to our case, incriminating for DuPont—in the sea of corporate secrecy. If the questions were poorly thought out or poorly crafted, the holes in the mesh would grow large, allowing key parts of the catch to escape. I even wrote a letter to 3M advising it of the lawsuit and putting it on notice not to destroy any existing data on PFOA. I scoured DuPont's corporate structure to determine who the key players were in the contamination saga and sent out deposition requests.

It was like the opening days after a declaration of war, when both sides scramble to prepare their troops and stockpile munitions. And then, when the first bullets fly, all hell breaks loose.

The moment our filing had been recorded in the court docket, DuPont's lawyers, as we had suspected they would, asked for a change of venue to Wood County, DuPont's domain. It was, after all, where the plant that was spewing out the PFOA was located. I countered that Du-Pont had operations all across the state, including Kanawha County, where we filed. But we had also sued the Lubeck Public Service District because it had colluded with DuPont in the letter sent to its customers claiming that their drinking water was perfectly safe. Lubeck was in Wood County and Wood County alone. The Service District attorneys argued that unless we wanted to dismiss them from the suit, the case would have to be assigned to Wood County. I'd never really expected to win that fight, but figured it was worth a try. The motion to change venue was granted; first skirmish to DuPont.

This and a mountain of other motions came my way, designed to keep litigation in limbo while DuPont prepared to deploy the tactic I'd been expecting them to use to try to kill any legal action in its infancy. They were stalling, and I knew exactly what was going on: they were going to try to use their new and not yet inked consent order with the regulators as a reason to have the suit thrown out. But I'd outmaneuvered them. Yes, they could argue that a consent order should quash any citizens' suit I might have filed in federal court, but my reading of the law was that such a consent order would not block us from pursuing a state common-law tort. Our new medical-monitoring claim, one that DuPont probably didn't think we'd be using in a class action, would be filed in state court.

After we filed, DuPont must have realized they were in trouble, but by then it was too late. By the time I filed the class action suit, the consent order was being finalized. Publicly announced just a few months later, the new consent order between DuPont and the state would follow the Cattle Team playbook and create two new teams of scientists, including scientists from DuPont: (1) a C8 Assessment of Toxicity Team (CAT Team) that would develop a new, government-approved screening level for PFOA in drinking water; and (2) a Groundwater Investigation Steering Team (GIST) that would oversee a massive new sampling program for PFOA in drinking-water supplies up and down the Ohio River. So, just as I'd hoped, DuPont's rush to try to thwart my threatened citizens' suit claims meant that they would no longer have a stranglehold on the water data. The regulators would now be intimately involved and reviewing DuPont's data.

And also just as I'd expected, when DuPont and the state publicly announced the signing of the new consent order in November 2001, the company's lawyers asked the state court judge to dismiss our class action suit, arguing that the case should be thrown out now that the agencies were looking into whether there was really any problem with the water. It sounded good, but they were shooting blanks. As I had hoped he would, the judge denied DuPont's motion, noting that such a consent order didn't have any impact on the medical-monitoring and other state common-law tort claims we were pursuing.

The idea that DuPont thought I could be persuaded that they were collaborating with government scientists in a neutral investigation, and therefore there was no need for either side to waste time and money on discovery, was ridiculous. I had heard that song before. There was no way in hell I was going to dance to it a second time.

I went on the offensive and forced the issue as quickly as possible by asking the court to impose firm deadlines for DuPont to turn over their documents. No more stalling. No more delays. Once again, the judge ruled in my favor.

When DuPont had asked for our case to be transferred to Wood County, they had gotten what they'd asked for, but probably not what they wanted. It was clear from the first ruling that the state court judge assigned to the new case, George Hill, was not following the DuPont game plan. Now in his seventies, Judge Hill was a highly principled judge and the kind of guy who excelled at everything he did. Born in Fairmont, West Virginia, he had attended Yale, where he was a halfback on the football team and a track star who set a Connecticut state record in high hurdles. After law school at West Virginia University, where he was the editor of the Law Review, he had served in the Korean War as a lieutenant commander on a navy destroyer. Judge Hill always came to court well prepared and had zero patience for lawyers who didn't. When we arrived for oral arguments, he had clearly studied the briefs and looked up relevant case law, which he could cite verbatim. He also had a BS radar the likes of which I'd never encountered.

Judge Hill rejected all DuPont's attempts to stop or delay our case. In January 2002, he ordered the company to produce the additional documents I was requesting by month's end. DuPont blew that deadline, as they had many others. My patience was gone. Enough was enough. I did something I had never done before: I filed a motion for sanctions. If granted, DuPont could be ordered to pay our attorneys' fees for all the time we had spent attempting to force their compliance.

That's when DuPont brought in the big guns. Outside lawyers from Steptoe & Johnson, a national law firm known for sophisticated litigation defense work, began appearing in court alongside Larry Winter's former colleagues from Spilman. Steve Fennell, from Steptoe's DC office, was brought in to oversee DuPont's massive task of document production. Larry Janssen, Fennell's colleague from the LA office, would become the new "face" of DuPont during court proceedings before Judge Hill, along with his young associate Libby Stennes.

I would be seeing a lot of Janssen. He fit the big-shot corporate attorney mold perfectly. Tall and trim, with a full head of gray hair and expensive suits, Janssen was adept at defending large corporations in bet-the-company lawsuits. His niche was mass torts and class actions. I admired his dignified style. Once he entered Judge Hill's courtroom, he didn't work the room the way some attorneys did; he sat quietly beside his co-counsel at the defense table, looking down at his notes, flipping a page over from time to time. When called on to respond to a motion, he'd scoot back his chair, stand up with perfect posture, button his top coat button, push back his glasses, and then speak calmly and directly. If he was losing an argument, he didn't pointlessly prolong the debate or put on a show of outrage or anger. He just sat back down and quietly scribbled a note on his legal pad. He never let anyone forget that he was representing big corporations and prominent businesspeople with all appropriate dignity.

It seemed to me that there was one other point he wanted to make with his comportment. Though he looked at others directly, he avoided all eye contact with me. If he had something to say, he'd walk up to the plaintiffs' table and address one of my co-counsels, never so much as glancing my way. It may have been all in my head, but I got the sense that he disapproved of my "switching sides" and joining the plaintiffs' lawyers he had been battling for decades.

. . .

As DuPont doubled down on their delay tactics, I continued to push back. I wanted DuPont to see that for every action on their part, there would be a forceful reaction on mine. When DuPont hauled us back to court on February 1, 2002, for a hearing in yet another attempt to stay discovery and stall the case, I took advantage of my in-person appearance before the judge to ask the court to force the case forward by setting a hearing date to decide whether to certify the case as a class action suit. DuPont objected, of course. The reason they gave for their objection made me shake my head: they hadn't had time to do discovery.

Judge Hill shared my dismay. He pointed out that five months had passed and DuPont hadn't bothered to even begin discovery. He set the class-certification hearing for the following month, March 22.

Class certification was a procedural but important hurdle. A class action suit provided a way for tens of thousands of people who were all exposed to the same chemical in their drinking water, caused by the same actions of DuPont, to resolve all their common factual and legal issues in one lawsuit, rather than having each person litigate individually. That would have meant repeating essentially the same case ten thousand times. The first step in this sort of case is to ask the court to certify your "class." If we got certification, the primary relief our class was seeking was clean drinking water and a class-wide medical-monitoring program.

. . .

Our proposed class was defined as everyone whose drinking water was contaminated with PFOA originating from Washington Works. We didn't know exactly how many people that was, but at a minimum, it included thousands of people on both sides of the Ohio River.

The defense typically challenges certification by arguing that the individuals and their claims vary too much from person to person to be considered common enough to be tried at the same time in one case. Our basic argument would be that all our class members had fundamental things in common: they all had the same chemical in their water, they all wanted it out, and they all wanted to get medical monitoring for their exposure. One surefire way to make the class certification process as complicated, time consuming, and expensive as possible is to bring in lots of rival expert witnesses. That's exactly the route DuPont wanted to go. Their lawyers told Judge Hill that they needed to bring in a phalanx of experts, arguing that this was necessary because the need for medical monitoring would necessarily vary by the individual, as each person would have unique vulnerabilities, different existing medical conditions, varying amounts of PFOA in his or her drinking water, and unequal levels of consumption of that water over different periods of time, leading to a different total "dose" of PFOA in each person in the class.

Judge Hill short-circuited that strategy. "I'm not going to allow dueling experts," he said. "Parties can submit briefs, and we'll have a hearing." I glanced at the lawyers at the defense table and could almost see the stomach acid boiling in their bellies. I felt certain that this was not how they'd imagined it would go.

We submitted our briefs and made our arguments—no experts in sight—and the judge ruled quickly: class certified. The case would now proceed as a class action for everyone with PFOA-contaminated drinking water attributable to the Washington Works plant.

Meanwhile, as DuPont was being forced to conduct more water sampling under the watch of regulators, the number of members in our class seemed to be growing by the day. Part of the class-certification order officially designated me and my West Virginia co-counsel as "class counsel," so we now officially represented and spoke for all these thousands of people. In addition to certifying the class, the court ruled that DuPont's documents would need to be produced. No more delays would be tolerated by Judge Hill.

A major hurdle was behind us. With rulings going in our direction, it began looking as though I hadn't led our firm down the rabbit hole to hell after all.

14

PRIVILEGED

February 11, 2002
Vincent, Ohio

Just before I became the newly certified counsel of our class of as yet uncounted thousands, I had the opportunity to meet some of the folks I was seeking to represent on the Ohio side of the river. Up to this point, only the Lubeck well field was publicly known to be contaminated with PFOA. When news reports appeared that the recent sampling by DuPont under the new consent order had found 7 ppb PFOA in the public water of Little Hocking, Ohio, the local water association called a public meeting. Representatives from the West Virginia DEP would be there, as would DuPont executives and Ohio EPA staff. I knew from my discovery documents that the company had found PFOA in Little Hocking's water as far back as 1984. But this was the first time the water district (or the people drinking its water) was hearing about it.

This I had to see. I was also eager for the chance to meet members of our proposed new class face-to-face.

On the day of the public meeting, 850 people crammed into the Warren High School auditorium in Little Hocking. Every one of the tan auditorium chairs was filled, and the folks who came late had to stand in the back and in the aisles down the sides. It was warm for February, and all those bodies crowded together pushed the temperature higher until the room was as hot as the discussion.

Up on the stage, Robert Griffin, the manager of the Little Hocking Water Association, introduced himself and the officials from DuPont and the West Virginia DEP who were sitting in folding chairs facing the crowd. I paid particular attention to DuPont's emissary, Washington Works plant manager Paul Bossert, a pleasant-looking, round-cheeked man of about forty. Dressed in a camel-colored sports coat with an open-collared white shirt that swelled around the curve of a comfortable belly, he had come armed with the company's preapproved talking points. The air was prickly with tension as Griffin gave a brief overview of the PFOA situation and the fact that PFOA had been found in the water. Then the questions came hard and fast.

A man rose from the audience and announced that he had done some online research. He said he had found a report that mentioned PFOA health effects such as tumors in fish gills, measurable changes in liver function, and birth defects.

"We know from this report that the chemical manufacturers knew the stuff was going into the water," he said. "I have a feeling the public is not being informed about what everybody standing down there knows. It's been in our water. If you guys knew it, we want some answers!"

The audience erupted with shouts.

"Are you going to answer the question?"

"Enough of this BS. Cut to the chase."

"We got kids drinking water around here!"

"Our lives are at stake here!"

The outbursts prompted cheers and applause.

Griffin tried to calm the crowd down. "We're upset as well. We got to do this in order, okay? They will answer your questions."

"They've already lied to us," someone replied.

"Enough of this crap!"

Griffin suggested a five-minute break before DuPont gave their presentation. The audience would not have it.

"We want to hear it now!"

"You don't get a break. We want to hear the answer!"

Then someone raised a very salient point: "I have a question directed to the water company. When they found this chemical, an unregulated chemical that could, of course, be dangerous to humans, why wasn't one of us notified immediately?" the speaker, who didn't introduce himself, said. "You sent out those little postcards about the meeting today. Why wasn't one of those postcards sent to us [saying] we don't know if it's dangerous? Maybe we [could] make our own choice about whether to let our children drink the water or whether to drink it ourselves. How long was it that the water company knew before you actually notified—wait, you didn't really notify anybody anyway. I mean, we saw it on the news."

Griffin replied, "We know that DuPont's been putting it in the river for apparently fifty years. We didn't know anything about PFOA in our water until the fifteenth of January [2002, about four weeks before the meeting], when we got the test results. They were brought to us by DuPont and the West Virginia Department of Environmental Protection. On that date we issued a news release that stated that."

"Fifty years?" a member of the audience said. "Why was it just now being test[ed]?"

"We knew DuPont was discharging it," Griffin said. "We did not know it was in our water."

"Sir," someone said, "enough of us have been here that we remember DDT. We were shown films of people being sprayed [with DDT] on the beach [saying] DDT is safe. You know, I think everybody is getting fed up with this. We're not blaming you, but let DuPont get up and say something, for crying out loud."

As I watched the drama escalate, it became clear to me that not only were these people angry, they were smart. They were not going to accept every answer at face value. They were doing their own research. I wondered if DuPont had ever dreamed it would go like this.

Finally Bossert was given the floor. He stood and took a few steps toward the audience. I think he was trying to communicate openness—nothing to hide here!—but I found his manner a little patronizing. He launched into a little homily about the company's "commitment to safety and health and the environment," then gave a presentation about PFOA. He acknowledged that it was biopersistent, which, he said, meant that "when exposed to it, it resides in the body for some period of time before it flushes out. That doesn't mean it does anything bad. It just means that it resides in the body for a little bit when you're exposed to it."

The use of such sugar-coated language to belittle the threat infuriated me.

It was all I could do to keep from shouting, but at that point I wanted to hear how DuPont would spin the story, so I held my tongue and just listened.

Bossert continued with a string of assertions—based on what, he didn't say: "It's not a developmental toxin. It's not a reproductive toxin. It's not a genotoxin, which means it doesn't affect reproduction. It doesn't affect childbirth. It doesn't affect the human genome, okay?"

He went on to speak of the plant's commitment, as part of the terms under the new state consent order, to reduce PFOA emissions by 50 percent by 2003. He cited the $15 million DuPont had spent since 1988 on "containment facilities" and their programs for "control, reclamation, and destruction" of the chemical, programs he said had already "reduced our emission by 75 percent." He said another $9 million had been allocated to reduce emissions by 90 percent by 2004.

"We've been using this material for over fifty years now," he said. "We've studied it at workplaces and in the community. We have nothing that would indicate there's anything to worry about. No human health effects."

A woman in the crowd shouted, "If there are no health hazards from this contaminant, then why are you spending millions of dollars to reduce the emission of it?"

Bossert, poker faced, responded, "We consistently try to reduce our environmental impact. That's one of the core values."

But this was a "chemical that invades your body," someone in the audience said. "You breathe it in. It's in the air. It's on your skin. You're drinking it in the water. You're probably eating it in your food because it contaminates the soil. It has to have an effect of some kind. I'd like to see the results of the studies."

Bossert promised to make them available.

A bald head rose from the crowd. "The reason I have no hair is because I have to go to chemotherapy. . . . Something caused my cancer. Something caused my sheep to die. Something caused my rabbits to die."

Bossert just kept repeating the company line: *There is no evidence of adverse human health effects.* But the crowd wasn't buying it. At one point, his cool deserted him and it appeared to me that he went off his talking points and said something unscripted and genuine: "I wouldn't be standing here talking about it if I had a belief there was something wrong!"

I couldn't help but wonder: Could he be telling the truth? Was it possible that the plant manager had been kept in the dark about the research his own company had done into the effects of PFOA?

Near the end of the meeting, I stood to speak. After identifying myself as an attorney handling the new case against DuPont, I directed my question to the non-DuPont officials. "We have not seen any reference to the 1 part per billion standard that DuPont has used internally for ten years for [PFOA], and I was wondering what the reason was for not referencing that standard."

Bossert took it upon himself to answer. "That's a maximum guideline," he said. "That's not a screening—"

A voice from the crowd, vibrating with anger, cut him off. "You don't want your employees to exceed that guideline, but it's okay for the community to be above that guideline?"

The Water Association people, state officials, and DuPont execs all sat there looking as though they'd just eaten something that didn't agree with them. Not one of them attempted to answer.

· · ·

Spring 2002
Cincinnati, Ohio

On April 19, a deceptively tidy package containing thirty-two computer disks arrived at our offices. I say deceptively tidy because those disks contained 248,000 pages of new documents from DuPont that, in the recently departed twentieth century, would have been wheeled into the building in a long train of luggage carts containing literally a ton of paper. Those disks were the fruit of my long battle against DuPont over my discovery requests, which they had been avoiding for nearly a year at this point. They had finally yielded because I had seen to it that their legal team had the threat of sanctions hanging over them. The judge had put his foot down. He had given DuPont an extended deadline to deliver but made it clear that if they missed the new deadline, my request for sanctions would be imposed. DuPont had had no option but to comply. Even now, I wasn't convinced this huge document dump contained everything I had requested. I hoped it would reveal a lot more about the extent of PFOA contamination, its threat to human health, and when DuPont had known about all that, but that remained to be seen.

I dug in once again. A large part of this latest batch consisted of something relatively new: email. Discovery now required companies to preserve and produce electronic documents. In those early days, many employees didn't realize that messages they thumbed on their BlackBerrys or typed on company laptops were stored on corporate servers, and they tended to be less guarded than they might have been in a memo or a more formal document. The proliferation of the devices, plus the ease of generating email, triggered an astronomical spike in the number of documents that had to be reviewed for discovery. Hence the ever-expanding army of attorneys DuPont hired to soldier through the process. For me, it meant a forest of wood pulp. I hated reading documents on a screen. For some reason ephemeral pixels blinking on a screen didn't stick in my brain, so I printed everything out. I needed to see it on paper, even if it took a ton of the stuff.

Once again, I noticed big gaps. Where were the records from Washington Works' sister plants in the Netherlands and Japan that, I had learned, were using PFOA? I had asked for all documents related to PFOA—not just those related to Washington Works—but DuPont was still withholding tens of thousands of other documents. I wasn't guessing. I knew they were because, as part of the discovery protocol, the company had had to send me "privilege logs," itemized lists of documents being withheld because their lawyers had decided to classify them as "privileged." Most legitimately privileged documents contain either specific types of protected communications between lawyers and their clients—typically, expressions of legal advice—or lawyer work product.

DuPont's privilege logs were the size of an unabridged dictionary. I had compiled plenty of privilege logs myself in my corporate defense role, and my gut told me that there was no way that much was privileged. I could dispute the privileged status of certain documents, but I'd have to do so one by one. DuPont would then have to give me the documents or prove to a judge that they deserved privilege. But who had time to read through hundreds of pages of lists? Most lawyers chucked those things in a corner or used them as plant stands. I assumed that this was what DuPont was counting on.

But I was part of the 1 percent who not only read them but scoured them. I took a deep breath, opened a fresh box of highlighters, and poured another cup of weak coffee—my preferred brew. Fighting with DuPont over privilege logs would be another slow-motion battle in a long war, but the privilege-logs fight would turn out to be one of the most pivotal moves in the whole class action. What I found in those logs would change everything.

In the meantime, there were plenty of unprivileged documents on those thirty-two disks that were eye-opening in their own right, including an email Bernie Reilly had sent to his grown son right after I'd called and told him I'd figured out that the Tennant case wasn't about cows, it was about Teflon. I hadn't stayed on the line long enough to let Bernie respond; I had delivered my message and hung up. Now, thanks to discovery, I was getting to see what his real-time reaction to that call had been.

"The shit is about to hit the fan in WV," he wrote to his son. "The lawyer for the farmer family finally realizes the surfactant issue. . . . Fuck him."

I laughed out loud, then grabbed the printout and almost ran down the hall to show Kathleen, Kim, and Tom in succession, leaving them muttering in amazement and shaking their heads at such an unvarnished look inside an opposing attorney's head. More important, it was further documentary proof that DuPont had known that PFOA was a problem all along.

At first I thought that DuPont's discovery fulfillment team must have included that email by mistake in the flood of tens of thousands of other emails. But it turned out to be just the first of several personal messages in which Bernie had talked a little too much shop to his son and family friends on his company computer.

Most attorney documents and emails, with rare exceptions, are typically protected by attorney-client privilege. Under normal circumstances, they shouldn't be accessible to opposing counsel. How could DuPont, which had been fighting us so hard on document production and so generous in designating things as "privileged," have let those slip through? But then I realized that since these were emails to Bernie's son and friends, not to a client or legal colleague, they were not protected under the privilege designation. Recovered from company computer servers because they contained the search words I had specified, they were all fair game.

It felt a bit strange to be reading Bernie's personal emails. But the messages served an important purpose in my plan of attack: they revealed what he (and therefore DuPont) had known at the time he wrote them, beginning in October 1998—the month Earl had first called me for help with his dying cows. Shortly after we had filed suit for the Tennants, as the Cattle Team was getting ready to deny that any problem chemicals had leaked from DuPont's landfill and instead blame all the dead cows on Earl, Bernie had revealed to his son that he'd known about the PFOA problem all along and was concerned that we'd find out about it. "[I] fly to Parkersburg tomorrow," he wrote. "Another long meeting to describe to the plant folks why the guy suing us over his cattle grazing downstream of our landfill would crucify us before a jury. . . . We really should not let situations arise like this, we should have used a commercial landfill and let them deal with these issues. Instead, the plant tries to save some money and apparently did not consider how it might look that this guy's cows are drinking the rainwater that has percolated through our waste."

. . .

Bernie had already known damn well that PFOA might kill animals. In November 1999, a few months before the misleading Cattle Team report came out, he had written to his son, "Parkersburg must announce to employees the results of the monkey study"—the one showing that when dosed with PFOA, the poor animals were dying miserable deaths. I found the announcement in a memo to plant workers in the same discovery dump. Thanks to the art of spin, I doubt that the announcement had generated much interest or concern among employees. The bulk of the memo had assured employees that existing procedures for handling PFOA were safe and protective of human health. The bad news about the two out of twenty-two monkeys that had died had been buried at the end with a comment that the deaths were "hard to understand."

Bernie's out-of-school comments made it crystal clear that despite their denials, DuPont had known—or at least suspected—all along what was causing the Tennants' problems. And I was increasingly confident that I could prove that they had taken very strategic steps to hide it.

15

ALTERNATIVE DATA

May 2002
Cincinnati, Ohio

Even though I'd been able to push DuPont into doing their water test-ing under the gaze of state regulators with the consent agreement, I was nervous about what results the so-called CAT Team set up under that agreement was going to produce. After all, this was the group charged with coming up with a new screening level for PFOA exposure in drink-ing water. It had been presented to the public as a group of unbiased scientists, but it included Gerry Kennedy, DuPont's lead toxicologist on PFOA matters, and John Whysner, hired by DuPont as an outside con-sultant, both men on DuPont's payroll. How unbiased could they be? To me, the CAT Team sounded—and smelled—a lot like the Cattle Team that DuPont had set up with the feds in the Tennant case.

We had already gotten a good idea of where the state's sympathies lay back in December, when the West Virginia Department of Environmen-tal Protection had held a meeting to assure the public it would figure out a proper safety level for PFOA. WVDEP scientific adviser Dr. Dee Ann Staats—the head of the CAT Team—tried to allay the public outcry over studies showing that PFOA had caused malignant tumors in rats. Clearly falling into step with DuPont's party line, Staats said that the rat tumors had been related to a metabolic process unique to rodents and therefore had no relevance to humans. Hogwash.

I was appalled that Dr. Staats was offering bromides to the public, suggesting that the troubling PFOA lab results in rodents were species specific, when I had seen data that proved otherwise, including testing on monkeys. This was the head of the CAT Team, for God's sake. Given that the community water supplies were clearly exceeding the old 1 ppb guideline, the stakes for getting this right couldn't have been higher. Under the terms of the consent agreement, DuPont would sample water within a one-mile radius of the plant. If PFOA was detected, the testing radius would expand by one mile. DuPont would be required to pay for an expensive filtration system to provide clean drinking water *only if* the new sample results showed PFOA above the new level established by the CAT Team—whatever that turned out to be.

Even as the trees in Cincinnati's Eden Park began to blossom, I couldn't ignore the dark cloud that was the CAT Team looming on the horizon. My belief in the incorruptibility of science was long gone. I'd seen government and private sector "cooperation" on such issues before, and I knew the result could be ugly—a testament to corporate interest and influence rather than scientific truth.

The CAT Team was scheduled to announce its "screening level" in May. While Ed and Harry stayed in touch with our thirteen named plaintiffs, answering questions and prepping them for possible depositions, Larry and I initiated a counterintelligence operation waged through the cumbersome and laborious filing of more Freedom of Information Act requests and old-fashioned dig-through-the-files investigation to try to keep tabs on the CAT Team's methodology and progress. If it served up bad science that favored the opposition, it would be an immediate disaster for the safety guideline but would also haunt our case throughout the litigation. I would have to become expert enough in the science to be able to recognize what was bogus and challenge it with the truth.

In addition to my already crushing workload, I took a crash course in the advanced sciences of water testing, toxicology, and risk assessment.

My thesis adviser in all of this was Dr. David Gray, PhD, of Sciences International, a consulting firm located outside Washington, DC. A toxicologist and risk assessor, David was now working as a consultant for our side, and we talked virtually every day, usually multiple times. To know if the number the CAT Team eventually came up with was reasonable, I would have to be able to understand the opaque science that turned tons of complex raw data into a single number purporting to represent how much PFOA was "safe" to drink. David helped me figure out what kind of documents I needed to be searching for in discovery and how to translate the inscrutable hieroglyphics into something meaningful and, hopefully, actionable. He also opened my eyes to the arcane science of risk assessment, which in the wrong hands could be manipulated to produce results that looked solid but were in fact a reflection of an agenda rather than reality.

Around that same time, the results of the new rounds of water tests required by the consent order from Lubeck, Little Hocking, and other sites downriver from the plant continued to come in. Many were exceeding 1 ppb. A few wells in Little Hocking were now showing levels of PFOA above 10 ppb.

As evidence of widening PFOA water contamination continued to mount and the levels of contamination seemed to escalate, I wrote more letters to the federal EPA asserting the community's urgent need of a source of clean drinking water. Residents, I said, shouldn't have to wait while DuPont and the state dithered. Exceeding DuPont's own safety guideline by a factor of ten should have been reason enough for EPA to act. The agency never responded directly to my letters. One test well in Little Hocking eventually came back at 35 ppb, *3,500 percent* higher than DuPont's original guideline. Local residents were understandably outraged. The newspaper reported a spike in bottled water sales. I watched all that unfold with building anger and frustration.

At least another prong of my strategy—making the facts about PFOA available to the public—was finally beginning to bear fruit. I had been including copies of key internal documents with our public court filings and continued submitting key documents to EPA and the state through letters warning of the pending public health threat. Finally, reporters outside Parkersburg had started picking up the PFOA story. The *Charleston Gazette* was the first out-of-town paper to run an in-depth exposé by investigative reporter Ken Ward, who used many of the documents made public by my court filings and letters to the regulators. A headline in the state's biggest paper raised awareness and alarm to new levels: "Wood Water Woes May Be Worse: Contamination May Be More Harmful than DuPont Says, Records Show."

Ward's story marked the debut of a new public voice for DuPont: Dr. Robert (Bobby) Rickard. A tall, fit, balding man, Dr. Rickard spoke with an air of scientific authority and confidence and carried the impressive title of Director of DuPont's Haskell Laboratory of Industrial Toxicology. In Ward's story, Rickard sang the company line that seemed to be DuPont's unwavering refrain:

"We've never had any adverse health effects from PFOA."

• • •

On Friday, May 10, 2002, I had just settled into the desk chair in my office when Harry faxed me a copy of an article from the *Parkersburg News and Sentinel*. The headline jumped up and bit me: "PFOA Safety Limits Set by Experts: DuPont Levels Well Below Parameters."

The story announced the CAT Team's new, government-approved safety level for PFOA in drinking water, which it had pointedly not shared with me before going public:

150 parts per billion.

I nearly fell out of my chair.

Surely there was some mistake. I went back and reread the number over and over, looking for a missing decimal point. Maybe it was supposed to say 150 parts per *trillion*, which would be 0.15 parts per *billion*. But it was no mistake. How could it be 150 times higher than DuPont's existing community guideline? This new number had no precedent in the thirty years of DuPont's own data or recommendations that I'd studied. It was orders of magnitude greater than anything that could have been remotely considered reasonable. How in the world had the CAT Team calculated that absurdly high number? I suspected that DuPont had seized control of the whole process and produced the result that would relieve them of any responsibility for fouling the community's water supply and cleaning it up.

Water below that level "would not cause harm to humans," Dr. Dee Ann Staats, the state toxicologist serving as the figurehead of the CAT Team, told the *Parkersburg News and Sentinel*. She was the same state government toxicologist who had parroted DuPont's "no harmful effects" line at the Lubeck public meeting. "I am confident the human protective levels established by the team are supported by the data," she continued. "These findings should bring comfort to the citizens of the Wood and Mason Counties."

DuPont's Paul Bossert echoed her reassuring message, saying that the CAT Team findings "support DuPont's position that the presence of PFOA at the low levels detected to date in drinking water in the Mid–Ohio Valley is not harmful."

All I could think of was what the Cattle Team had told Earl three years earlier: that there were no signs that DuPont's chemicals were causing health issues; that it was his fault his cattle were dying.

Once again, DuPont's company line had the cover of science, cosigned by the state DEP. Bossert could insist with a straight face that the water in the valley was perfectly safe because every single sample tested so far—even the high results in Little Hocking—was now well below the new safety level. With the announcement of this new guideline, the threat had gone from alarming to innocuous overnight. And my chances of prevailing in our class action had been nearly extinguished—if contamination below 150 ppb was declared "safe" by government scientists, what case did I have?

On the surface, the CAT Team report was a disaster. But I knew I had to resist the despair that tempted me. I reminded myself that in reality, nothing had changed. The *actual* levels of PFOA in the water had not changed. Neither had the fact that people were drinking it. The only thing that had changed was the measuring stick. It was now up to me to prove that the stick was warped.

. . .

Long before she became my wife, Sarah would frequently tell me that I'm
naturally suspicious of everything. I'm always reading between the lines,
looking for the hidden catch. We're taught that studies yield data. Data
doesn't lie. But not all data is created equal. Sometimes there's a flaw in
the method. The measuring tool might be broken. Or perhaps it's measuring the wrong thing. Scientific conclusions can be skewed by subconscious
bias or honest human error. Sometimes honesty has nothing to do with
it; data can be manipulated. Scientific findings can be reverse engineered.

Ever since I had deposed Andrew Hartten and seemed to hit such a
sore point when I had probed DuPont's decision to switch water-testing
labs, I had been trying to figure out what was going on, what nerve I had
struck. One of Bernie's emails had mentioned a "new test protocol," and
he worried that it could show "much larger" concentrations of PFOA in
drinking water. At the time he wrote that email, DuPont had been telling
the public that the PFOA levels in local water were safe. The October
2000 letter to the Lubeck public water customers—the one that had landed
in Joe Kiger's mailbox—had claimed that the level of PFOA found in the
Lubeck water in August 2000 was comfortably below 1 ppb. If that was
true, it meant that the level had somehow dropped significantly since the
2 to 3 ppb levels I had seen in DuPont's internal files from the 1990s. If
PFOA was slowly leaving the water system, why was DuPont so concerned
that new test protocols might show much larger concentrations?

To get to the bottom of this mystery, I turned to the expert who had
first helped me see the connection between PFOA and PFOS. After reviewing and studying all the raw data I sent him, he called to explain the
story he'd pieced together. He carefully explained how DuPont had been
generating internal documents since the early 1990s that made the PFOA
contamination look only half as bad as it really was. It seems DuPont had
switched to an outside lab for their water testing, after their own lab had
found levels far above 1 ppb in the local water supply as late as 1991. The
new, outside lab that had been testing the water for the company since
1991 had generated reports with two parts: the first consisting of raw

data, the second an analysis indicating that the limitations of its testing method meant that the actual amount of PFOA in the water was substantially higher (about double) than it appeared to be in the raw data. Yet when DuPont eventually sent out the test results to the local water district, they sent only the raw data. The personnel in the small West Virginia water district weren't experts and didn't realize that something was missing, so to them it appeared that the new Lubeck wells had tested below the 1 ppb safety guideline. So an actual measurement of 1.6 ppb—well above the guideline—could magically become 0.8 ppb, just below it.

None of that data had been reviewed outside the company beyond the local water district. This was why it had been so important to force DuPont into the consent agreement that mandated supervised water testing. The government scientists now looking over DuPont's shoulder would likely see through the misleading raw data and realize that actual PFOA levels were likely much higher than the final numbers reported to the water district had suggested.

But even if DuPont could no longer get away with understating the actual level of PFOA contamination, they could *overstate* the level of "safe" contamination. That had led to their successfully setting up the CAT Team with its wildly inflated 150 ppb safety limit.

DuPont had won that round. It was too late for me to reverse the CAT Team's guideline number, but I could still ultimately discredit it—in court.

This is why I had spent hundreds of hours attempting to learn the science myself, instead of simply relying on predigested summaries. My study had provided the necessary base I needed to hunker down with my risk assessment expert, David Gray, and focus on finding false fronts and trapdoors in the supposedly objective scientific reasoning that had led to the clearly bogus result. We spent more hours on the phone as he walked me through the risk formulas and scientific calculations used to come up with safety guidelines. Dr. Gray had spent decades in the field of toxicology, and his work had helped create and shape many of the methods and formulas now in use.

Up until this point in my career, I had worked only with the end point of risk assessments—the guideline number itself. Now I was learning that the ostensibly scientific process involved a surprising degree of subjectivity. Certain variables known as "safety factors" or "uncertainty factors" were plugged into the standardized risk assessment formulas to calculate a guideline. Each variable had to be assigned a value. That value was the personal decision of a scientist.

Making the wrong call on the value of one or more of the variables could change the outcome by a factor of ten, one hundred, one thousand—or more.

Independently, Dr. Gray had done his own risk assessment calculations for PFOA based on the data we had available. His number made the CAT Team's level seem even more suspicious. According to Dr. Gray, even the original DuPont assessment of 1 ppb had been far too high. His data could not support a safety level above 0.3 ppb, which was not even in the same galaxy as the CAT Team's 150 ppb.

I felt confident that Dr. Gray's number was closer to the truth, but his assessment alone wasn't enough. Ultimately, I would have to persuasively debunk the CAT Team's number, which was, after all, the product of a half-dozen proclaimed "experts," not just one. To do that, I had to know what had actually transpired during the CAT Team risk assessment meetings. How had the team come up with 150 ppb when looking at the same raw data Dr. Gray had reviewed? I decided that the first step in getting to the bottom of this question was to depose the state's leader of the CAT Team—Dr. Staats—and ask her myself, under oath and on the record.

16

APPETITE FOR DESTRUCTION

June 2002
Charleston, West Virginia

We met at Ed Hill's firm in Charleston. Ed and Larry Winter both sat in. A videographer aimed a camera at Dr. Staats, who sat at the end of a large conference table surrounded by attorneys from the state and DuPont's outside counsel, Steptoe and Spilman. DuPont's in-house attorney, John Bowman, had flown in from Wilmington. Throughout the interview that was about to take place, DuPont's attorneys would maintain unflinching, impassive poker faces.

Dr. Staats, fortyish, one-upped them. Her expression approximated a sheet of cold steel throughout the proceedings. She frequently flipped her shoulder-length brown hair with her hand, almost as if she were physically brushing back my questions as she might annoying insects. Dr. Gray sat next to me in case I needed help with toxicology lingo or risk-assessment translations. I thought a sensible place to start was to ask for access to Dr. Staats's notes. They would give us some insight into the methodology and variables used and the calculations. Here's how that went:

Did you take any notes?
Yes.

Do you have your notes?
No.
What happened to your notes?
I copied them, faxed them to TERA [an outside consulting firm on risk assessment that DuPont had recommended that the state retain to coordinate the CAT Team's work], made sure they had them, and destroyed them.

Had I heard that right? I glanced around at Larry and Ed and then tried to read the faces of DuPont's counsel. Inscrutable. Had a state employee just said, on the record, that she had destroyed official documents?

I followed up, asking Dr. Staats about her phone conversation with TERA consultants the day before this deposition.

Did you ask them to send to you your notes that you had faxed to them?
No.
Why not?
Because I knew they had already destroyed them.
That TERA had destroyed them?
Uh-huh.

Destroying documents in the middle of a lawsuit? The law had a name for this: "spoliation of evidence." In this instance, the judge could potentially order the witnesses to turn over their computers so we could attempt to retrieve the deleted materials.

Staats said that it was her standard practice to destroy notes from every meeting. It was simply "habit, out of years of doing litigation," she said. "I don't keep drafts. I don't keep notes. I don't keep emails."

Destroying notes is not completely outside the norm. It's standard practice at many firms to destroy records once they are no longer needed.

However, a state official destroying official records related to ongoing litigation did not fill that bill. As someone schooled by "years of doing litigation," this wasn't her first rodeo. She'd known we would be coming for her notes, a fact she proceeded to admit.

"I fully expected that you would subpoena me," she said, as if destroying records in the face of a subpoena were the most natural, legal thing in the world.

. . .

After the deposition, I swapped reactions with Ed and Larry. They were as taken aback as I was. Dr. Staats had been appointed as a state employee to participate in a governmental process outside of (but related to) ongoing litigation. Of course she could be subpoenaed in a lawsuit. But as a scientist and a government employee, the assumption—and the *obligation*—was that she would be neutral. Her deposition shattered what little illusion I had left that the state regulators were free of DuPont's influence. I wasn't just surprised, I was shocked. And furious. Who else was destroying records? Other regulatory officials? DuPont employees? We needed to act quickly to stop it. I filed a motion laying out for the judge the outrageousness of a state expert supposedly working to make a neutral, scientific determination about health risks to the public destroying documents she knew would be relevant to ongoing litigation. I asked for an injunction to prohibit the further destruction of documents and to force the state to turn over its computers. Our forensic computer experts could try to retrieve whatever had been deleted.

The court held an emergency hearing the very next day. Our team, the state's, and DuPont's crowded into Judge Hill's court. He didn't look happy.

"Does the department plan to continue destroying these documents if the injunction is not issued?" he asked Christopher Negley, the lawyer representing the state of West Virginia.

"We are going to do what we customarily do," Negley said.

That definitely wasn't a "no."

Negley argued that Dr. Staats had not violated any state policy by destroying her notes, because the state had no such policy. That decision was left to the discretion of each employee. Never mind that Staats had admitted that she expected to be subpoenaed after the CAT Team's meeting to set the new PFOA drinking water level; destroying notes was simply her "customary practice."

Judge Hill was not satisfied with that answer. "Does that make it not obstruction of justice, or what?" he asked.

"I believe, Your Honor, she has her right to do her business as she sees fit."

"So did Richard Nixon," the judge shot back.

I tried not to smile.

"I think it's a no-brainer," the judge said. "It is a crime, and I think it should be enjoined."

Injunction: granted.

DuPont appealed to the Supreme Court of Appeals of West Virginia, seeking to quash the injunction. The Supreme Court denied the appeal. Computers: seized and searched. Deleted files: recovered.

To no surprise of mine, one of the state's documents—probably one it really hadn't wanted us to see—revealed that Staats had come up with her own risk-assessment number before the CAT Team process had even begun. That number was 1 ppb.

The same day that West Virginia's highest court upheld the injunction, DuPont's lawyers disclosed that they had "just discovered" that DuPont's lead PFOA toxicologist, Gerry Kennedy, also on the CAT Team, had also been destroying documents—not just from the CAT Team meetings but possibly also some other communications concerning PFOA. Turned out I was still capable of being amazed: DuPont was destroying documents, too. Big mistake. If this had been an attempt to keep us in the dark about the risks of PFOA, not only would it fail, it would provide me with an opportunity to shift the balance in an eventual trial dramatically in our favor. I amended my motion for sanctions over the missed discovery deadlines. I argued that this new disclosure about Kennedy supported the need for severe sanctions, which should include the judge instructing a future jury to consider it a given—since we couldn't know all the documents Kennedy had destroyed—that PFOA was toxic to humans. Sometimes the best defense is a good offense.

In response, DuPont's lawyers claimed that Kennedy's actions had been an honest mistake. An inadvertent misunderstanding. It wasn't their fault. They had advised Kennedy to preserve and protect his PFOA documents for the purposes of this case. They painted him as someone getting up there in age who didn't like—or fully understand—the new world of email and electronic files. Kennedy, they said, had simply misunderstood their instructions.

In what I assumed was an attempt to bolster their position that the destruction of documents was a mistake and not part of some misbegotten legal strategy, DuPont's in-house attorneys filed affidavits with the court to set the record straight. The affidavits spelled out what they had—and had not—instructed their clients to do. They even attached copies of internal documents providing that advice. This was a huge calculated risk by DuPont that revealed how concerned they were that they might get sanctioned for document-destruction, given how steamed Judge Hill already was over the Dr. Staats document-destruction debacle. DuPont's lawyers had willingly shared attorney-client legal advice—the most typically bulletproof of "privileged" documents—the ones I'd never dreamed of getting.

But at what cost? By introducing the issue of what they, the lawyers, had told Kennedy to do and say about his records, they had waived any claim of attorney-client privilege over those communications and made all such instructions about document preservation, to Kennedy or any other DuPont employees, a matter I had a right to cross-examine, opening the door for me to demand similar normally privileged documents. And that was only part of the damage they had brought on themselves. By decoding the messages instructing DuPonters what to say and not say, write and not write, I was able to see exactly which subject areas they were most concerned about. I could also see in a convenient email header all the people they were concerned might know something that could damage their case.

These were just the people I wanted to depose.

17

OF MOUSE AND MAN

July 2002
Cincinnati, Ohio

With the class only recently certified, I anticipated a trial date—not yet formally set—sometime late in 2003. That was a year and a half off, which might sound like a long time, but on the legal calendar it was just around the corner. In addition to the exhaustive and exhausting research into technical issues, I was more than occupied fighting with DuPont over document production issues—which required an almost daily barrage of letters and notices—and trying to learn what had gone on inside the black box of the DuPont–West Virginia collaboration on the CAT Team. DuPont was making both endeavors as complicated, difficult, and time consuming as possible. It was not glamorous work. I sat in my office for ten, twelve hours a day reviewing documents and exchanging letters with DuPont's attorneys, occasionally punctuated by frustrating, sometimes hostile phone calls.

I was handling the research and discovery almost entirely by myself. At that point, it would have been impossible to bring my co-counsels up to speed on the background necessary to make sense of any of it— background it had taken me all of three years to accumulate. Even I had to rely on the help of my handpicked experts. Now I had to become fluent in the language of toxicology. What happens when safe levels are exceeded? That was the critical element of risk to human health that I would need to prove for our suit to succeed. There could be millions of gallons of PFOA in the water, but if it presented no risk to anyone, I would have no case.

DuPont scientists had been doing PFOA animal studies for more than half a century, including the lethal monkey studies, to answer that very question. So I'd always known there was plenty of data, which was why I had fought so long and hard—and was still fighting—to get DuPont to turn all of it over. Now that I was finally starting to get the studies into my hands, I needed time to decipher them. They would tell me exactly what DuPont knew about the toxicity of PFOA—and when they had known it.

I pulled out hundreds of documents—meeting notes, memos, lab reports, raw data, letters, study abstracts—that offered some salient clues. Even this whittled-down batch contained too much information to keep straight in my head at once. The only way I knew to process it was to order the pages chronologically and lay them all out on the floor with my usual color-coding and sticky-note categorization.

When I did that, my office floor became a sort of time machine. Each page in isolation offered a single snapshot, a fleeting moment in DuPont's long history. The documents revealed what the company had known—or hadn't—at a particular point in time. But when I put them together in succession, they merged like individual frames in a stuttering black-and-white film. The same characters entered the picture over and over.

It was clear that DuPont's second century had been founded upon science. DuPont had not been created to be some evil empire intent on polluting the world. To the contrary: for decades it had been grounded in the principles, methods, and ethics of science. Its researchers had been well-respected, contributing members of the greater scientific community. They had valued good experimental design, peer review, and publication. They had shared their findings with a professed faith in the advancement of knowledge through science. They had espoused a duty to report "the unvarnished facts," not only to other scientists but to customers, employees, and the public. And that ethos had come straight from the top. For thirty years, from the 1950s into the 1970s, top DuPont executives had promoted a strong ethic requiring any sign of chemical toxicity to be published.

"It all comes down to credibility" was the way Dr. John A. Zapp, Jr., the director of DuPont's research arm from 1952 to 1976, put it. "You've got to have credibility or you're wasting the company's money."

I had been piecing together the multilayered narrative buried in the documents ever since I took the Tennant case, and I would continue to develop it for many years. The process was tedious and exhausting—mentally, physically, and emotionally. It involved hours of crawling around on a paper-white floor. It was like archaeology, where fragmentary clues were unearthed glacially, with a brush and not a spade.

The story I eventually pieced together revealed a decades-long scientific saga unfolding within the corporation. There was no one person, event, or decision responsible for the presence of PFOA in the community's water and blood. It was the result of a cascade of decisions made by many people within DuPont, for many reasons, over many years.

By the time PFOA came along, DuPont's in-house scientific laboratory, commonly referred to internally as Haskell Labs, had been doing pioneering work in toxicological research for more than a decade as a part of DuPont's internal controls. By the late 1940s, Haskell had developed a routine sequence of animal tests to evaluate the harmful effects of a chemical on living creatures. Mice and rats are often the first to be used in studies because they are among the cheapest and have brief life spans, allowing researchers to study "lifelong" effects in a relatively short period of time. If there are any reasons for concern, the tests often progress to studies on more complex, longer-lived species, such as dogs and then monkeys, which are increasingly expensive. In 1950, an animal cancer study could cost up to $200,000, the equivalent of $1 million in today's dollars—half of Haskell Labs' annual budget at the time.

But DuPont had historically considered those studies a worthy investment—to protect their workers, the public, and, not least, their bottom line. Their scientists began investigating the toxicity of PFOA around 1954, just three years after the first shipments began arriving at Washington Works from 3M. By then DuPont already had enough cause for concern to warn workers about exposure to the chemical. A memo from that year I pulled out of my stacks advised handlers of the compound to "avoid excessive exposure of the skin" and to "not inhale the dust or fumes."

Animal testing of PFOA, however, did not begin in earnest until the 1960s. The head of the Toxicology Section warned that PFOA should be "handled with extreme care" and commenced studying the compound's toxicity in lab animals. The first studies did little to ease concerns.

One such study was a rat experiment in 1962, the year Teflon was approved for use in consumer cookware and began entering widespread circulation. The potential profits were huge. DuPont, historically cautious about rushing to market before thorough testing, was uncharacteristically racing to catch up with its nonstick-cookware competitors.

In the experiment, six lab rats were fed ten daily doses of PFOA for twelve consecutive days. The rats were then autopsied, some shortly after the tenth dose and the rest fourteen days later. Both groups showed moderate enlargement of the liver and slight enlargement of the testes, kidneys, and adrenals. Researchers noted that "cumulative liver, kidney, and pancreatic changes can be induced in young rats by relatively low doses of PFOA." Any changes to organs in these tests were a strong indication of toxicity and were frequently warning signs of future disease.

Three years later, in 1965, a dog study showed toxic liver damage in beagles exposed to PFOA. The simple fact that dogs were being tested, and not just rats, indicated the high level of concern at DuPont—dog tests being far more expensive and undertaken only when rat testing produced alarming results.

Compounding the bad news from DuPont's expanding animal studies, the PFOA problem became even thornier when, in 1972, the Marine Protection, Research, and Sanctuaries Act was passed, expanding the protection of marine life by regulating the disposal of chemicals into the sea. For years, DuPont had been dumping PFOA wastes into drums destined for the bottom of the ocean after realizing that solid waste from

Teflon dumped into landfills near their plant would leach into the groundwater. Now DuPont returned to land disposal, despite the leaching problem.

That combination—a known animal toxin leaching out of landfills into water—inevitably led to some unpleasant news shared in a 1978 meeting between DuPont and 3M. That was the year 3M informed Du-Pont that it had discovered PFOA in 3M workers' blood. After the meeting, the grim news was sent all the way up the chain of command and shared with DuPont medical director Dr. Bruce Karrh. He was also sent the now two-year-old paper on fluorine in blood by Guy and Taves revealing that organic fluorine was showing up in the general population, as indicated by tests of blood-bank samples across the United States. Guy and Taves had strongly suspected that was related to man-made, industrial fluorochemcials—chemicals such as PFOA. Karrh, forty-two years old, was an outspoken advocate of transparency in industrial medicine. Just a year earlier he had presented a paper of his own—"A Company's Duty to Report Health Hazards"—to a conference on ethical issues in occupational medicine. Very much in the DuPont mold of the 1950s and '60s, he contended that it was a company's responsibility to "be candid and lay all the facts on the table" and that that was "the only responsible and ethical way to go."

Dr. Karrh saw DuPont, the nation's biggest chemical firm, as being "in the eye of the public occupational health storm." In leading the industry through this storm, he claimed that DuPont aimed "to meet or go beyond the requirements" of laws and regulations. That included full disclosure of the "unvarnished facts about health hazards." Radical honesty was critical for maintaining credibility with employees, customers, and the public. "To do less," he wrote, "would be both morally irresponsible and, in many instances, economically damaging."

This philosophy matched Dr. Karrh's aggressive response to 3M's news: all employees involved in making Teflon would be briefed on the new blood findings, followed by a complete review of operations to ensure safety procedures were limiting exposure. Employee medical records would be reviewed and the blood of potentially exposed workers would be tested. To establish a baseline, the company would also draw blood from employees who purportedly had no risk of PFOA exposure in the workplace. If the physical exams and the blood tests showed any unusual trends, an epidemiological study would be considered.

The plan sprang into action at Washington Works, executed by Dr. Younger Lovelace Power, the plant's medical doctor, whose stated goal was "to make people live longer, better." Dr. Power reviewed the medical records of the eleven Teflon operators and eighteen laboratorians whom he identified as having potential exposure to PFOA.

As at 3M, those workers had elevated levels of PFOA in their blood. Even though Washington Works formally claimed to have found "no unusual health problems" in exposed workers, Power was "disturbed by the frequency of borderline elevated liver function."

In a similar study of employees at DuPont's Chambers Works plant in New Jersey in 1979, exposed workers showed a "notably higher" number of abnormal liver-function tests than did unexposed workers. Somehow, despite being "notably higher," the finding was deemed "not statistically significant." Nonetheless, DuPont volleyed those results back over to 3M with a note about "our general practice of reporting or otherwise publicizing relevant findings even if they are not required to be reported."

Yet despite all these warning signs—and all the talk of moral duty and a culture of transparency—once again, the company did not report their findings to EPA.

. . .

I periodically had to come up for air and clear my head. Where other guys might grab their buddies and head out for a few beers, I headed to a storage garage in Cincinnati where I kept my 1976 Buick Electra 225. Black, shiny, not a single door ding or scratch on it. Sarah teased that it was like floating on a big comfy couch, but I loved driving it. It was as close as I could come to the car my parents bought off the showroom floor in 1973, the one I helped my dad select. Just sitting in it took me to my comfort zone.

Then it was back to the office, where my paper pile continued to add to the disturbing timeline: After confirming the presence of fluorine in workers' blood in 1978, 3M and DuPont met with their lawyers to discuss a critical question: Should they tell EPA? The two-year-old Toxic Substances Control Act required them to report any "substantial risk" data on existing chemicals. Yet the companies believed there was a loophole that applied here. If there was published information on the matter already available to EPA, the companies argued, they were relieved of the need to report it. They decided that the Guy and Taves study—published and publicly available—gave them that out. Moreover, they could point to DuPont's research that purported to show that PFOA exposure at normal plant levels presented no "substantial risk" to workers. That conclusion, mind you, was "primarily based on the absence of any known health effects."

When I read this in a corporate memo, it gave me a major tilt. What about the liver effects in rats and dogs, coupled with the worker studies that also indicated concern about the liver? What evidence did DuPont have that the animal and human liver findings were unrelated? It didn't make sense. The company seemed to be ignoring one of the core principles of toxicology my experts had taught me: if a substance shows toxic effects in animals, you must assume it is likely toxic to humans until proven otherwise.

As DuPont was testing their workers' blood, 3M was conducting a ninety-day rat and primate study. The results came out in November 1978: adverse effects had been found in both rats and rhesus monkeys. The monkeys—which are more biologically similar to humans—had fared much worse than the rats. They had suffered gastrointestinal effects and signs of "hematopoietic effects," which indicated a problem with the body's ability to make blood and blood cells. Clinical signs of toxicity were evident even in the monkeys given the lowest dose. All the monkeys given the highest dose died within one month.

I scoured the stacks of studies but could not find evidence of another monkey study until the one performed in 1999—nearly twenty years later. What sense did it make to find monkeys keeling over in one study, then just forget about it for two decades?

It was then that I began to notice that DuPont executives repeated the same catchphrase frequently, even after the worrisome study results: they had "seen no evidence of adverse health effects" from PFOA in humans. The phrase appeared word for word throughout the timeline.

The bad news from animal testing kept coming. In 1979, a DuPont document summarized the effects of PFOA exposure in multiple species. Rats suffered liver degeneration, enlarged livers, and increased liver enzymes. Rabbits with skin exposure lost weight and exhibited labored breathing. Two dogs died within forty-eight hours.

The stakes rose higher in 1980, when DuPont confirmed that PFOA is bioaccumulative, meaning that trace amounts of exposure could build up in the body over time. This compounded the 1979 confirmation that PFOA was biopersistent—once it entered the body, it didn't break down. The decay rate was slow. The body did not know how to metabolize the man-made molecule. The compound showed a tendency to bind to albumen, a blood protein, causing it to linger in the blood and circulate throughout the body. These two qualities compounded each other, meaning that exposure to a very small amount of PFOA over a long period of time could lead to significant, increasing levels in the blood.

· · ·

The passage of the Comprehensive Environmental Response, Compensation, and Liability Act, otherwise known as the Superfund Act, that same year, 1980, was a key milestone in the saga. It ratcheted up the potential costs—in the form of massive liability—for corporations such as DuPont that generated a wake of hazardous substances. Now, for the first time, DuPont could be held accountable not only for future releases but for years of past contamination. But that applied only to regulated hazardous substances. PFOA, of course, was not regulated as a hazardous substance, so it would not trigger any liability under the Superfund Act. Companies that made or used PFOA had a considerable interest in keeping it unregulated. By now the plastics industry was a commercial and cultural juggernaut. The volume of plastics production had surpassed that of steel. It was an inconvenient time for the evidence of PFOA's potential hazards to be mounting. Though much of the worrying evidence was generated by DuPont's own scientists, those same scientists claimed that there was much they still did not know about PFOA. DuPont's assistant medical director noted the concern about the elevated liver enzymes—a sign of possible liver damage—in certain Teflon workers. "Even though we have found no 'conclusive evidence,'" he wrote, "we still cannot explain [it]." A draft of a corporate memo seemed to suggest that DuPont had decided not even to try to find an explanation, to the assistant medical director's consternation. "I am concerned," he wrote, "[that] the draft implies the medical division will not continue the study of liver tests on those employees exposed to PFOA."

Meanwhile, the narrative spread out all over my floor showed that though DuPont seemed to be shying away from doing more medical research into PFOA toxicity, at the very same time they were beginning to treat it as a possible hazard to their own employees. High-level meetings were called to address workers' exposure to PFOA. The result was a recommendation that handlers wear protective clothing, including, at minimum, gloves, breathing protection, and disposable clothing.

DuPont made changes in equipment and infrastructure as well. In the Teflon division, they brought the process of PFOA mixing under the exhaust hood and raised the air-supply inlets on the dryers to remove the PFOA-rich air that gathered under the ceiling. They sealed leaky seams in dryer doors and added inspection windows to reduce the need to open those doors more than necessary. They knew that PFOA could permeate all protective materials eventually, so all gloves had to be disposed after use. Workers with inhalation exposure were advised to wear respirators and protective masks.

All of these measures were announced in a PFOA communications meeting on July 31, 1980, to discuss the company's immediate and long-term plans in light of recent findings. DuPont officials described the chemical's toxicity, which varied depending on the pathway: orally, it was claimed to be "slightly toxic"; with skin exposure, "slightly to moderately toxic"; and inhaled, it was "highly toxic."

During the meeting, the DuPont executives went over the initial blood test results, both internally and from 3M. By now they had learned that organic fluorine levels in workers' blood "generally correlate with job exposure potential." This was a clear concern. Although the meeting notes again sounded the refrain that PFOA "has caused no health effects," they concluded somewhat contradictorily that "continued exposure is not tolerable."

Ridiculous. If something had zero health effects, why would exposure be considered intolerable? DuPont provided the answer on the very next page.

"After 25 years of handling PFOA we see no damage among the workers," the notes said. "However the potential is there—PFOA has accumulated in the blood."

Now I understood. What was of utmost concern to the DuPont scientists was the unusual biopersistence coupled with the bioaccumulation of PFOA. Every exposure—no matter how tiny—could enable more and more PFOA to build up in the blood and human body over time. Once present, it would stay there for years. That meant the chemical would remain in people well into the future, where it could present the potential for untold harms that might not manifest as a clinical disease for decades. In short, PFOA in the body was a ticking time bomb.

. . .

Just as DuPont was ramping up concern over the effects of PFOA on workers—and the associated possible financial liabilities—the chemical industry as a whole was entering a new era of accountability. The following year, 1981, marked the beginning of New York's Love Canal cleanup, one of the first and largest Superfund projects in history that attempted to rectify a quarter century of companies dumping toxic chemicals into inadequate landfills. Rachel Carson, the author of *Silent Spring*, was posthumously awarded the Presidential Medal of Freedom for alerting the world to the devastating environmental impact of uncontrolled pesticide use. As if DuPont were girding for the new era, Haskell Labs expanded again, adding thirty-nine thousand square feet of space for genetic toxicology, aquatic studies, industrial hygiene, biochemistry, and a library.

A dramatic development occurred later that year when a 3M rat study indicated that PFOA caused birth defects in unborn rats. The pregnant dams had been fed PFOA through stomach intubation. The rats had been sacrificed before giving birth, and the fetuses had been closely examined. They had revealed a consistent problem: eye defects. The results were too clear to be swept under the rug. As required by the Toxic Substances Control Act, 3M disclosed the study results to the government. On March 20, 1981, 3M shared the news with DuPont.

Evidence that PFOA might be related to birth defects sent DuPont spinning into an internal state of alarm. Seven days after receiving the news, scientists from DuPont made a visit to 3M to verify the study results. After confirming that the results were valid, DuPont spent several days preparing internal procedures and developing a communications plan to share the news with employees, including a Q and A with thirty-nine items anticipating workers' concerns. One of the items was:

Do you have any knowledge of DuPont employees . . . who have been exposed to PFOA whose children suffered birth defects?
We know of no evidence of birth defects caused by PFOA. . . . We will investigate further.

Alarm about the results was so high that on April 1, 1981, two weeks after receiving the 3M study, DuPont removed all female employees from Teflon-related jobs for reassignment and began sampling their blood. On April 6, Dr. Karrh, still DuPont's chief medical officer, resuscitated a worker pregnancy study that had been put on hold. Prepared by DuPont epidemiologist Dr. William Fayerweather, the study sought "to determine whether pregnancy outcome among female Washington Works employees is causally related to their occupational exposure to C8 [PFOA]."

That same day, a communiqué was issued to DuPont employees. The notice reported that the 3M study had found that PFOA fed to pregnant rats caused birth defects. "At this time, we do not know the significance, if any, of the preliminary animal experiment as it may relate to employee exposure," the notice said. "Further studies are planned to define possible reproductive effects."

Then, consistent with Fayerweather's study design, DuPont collected blood data and reviewed birth records for seven Teflon employees who had recently given birth. All had elevated PFOA levels in their blood. Two of the seven babies had defects noted at birth. Both of those were eye defects. Fayerweather had made it clear in his study outline that, in the general population, the expected rate of birth defects involving the eyes is two for every one thousand live births. DuPont had now found two in *seven* births.

This was no longer about rats. It was about humans.

DuPont never completed their pregnancy study or disclosed the results of their review—finding two out of seven human babies with eye defects— to governmental regulators. When a later rat study failed to find the same eye defects, DuPont pointed to it as proof that the initial 3M results had been flawed. 3M also went back to EPA and told it that the original rat birth-defect findings it had reported under TSCA were invalid and worthy of no further investigation. EPA apparently accepted 3M's claims at face value and looked no further into the issue. Even before the end of 1982, DuPont allowed their female employees to return to Teflon.

Though I knew each new find in my review of the documents was making our case stronger, it was personally harrowing to return to it day after day, learning about all the pain endured by the lab animals and the suffering of human mothers and their babies. It made me wonder: If reading about this years after the fact was difficult, how had it been for the scientists who had observed it happening right in front of them? They had reported it up the chain, yet everything had gone on as if nothing had happened. The company had made some changes in safety precautions, sure, had even taken the women out of the Teflon lab for a short period of time, but apparently no one had ever seriously considered ceasing to use the implicated chemical or reported any of the troubling human data to regulators.

It took another seven years before a fresh wave of panic rocked DuPont in 1988 with cancer results in a rat study. The new two-year cancer study had found that PFOA caused Leydig cell (testicular) tumors in rats. DuPont relied on this data to classify PFOA internally as a confirmed animal carcinogen and possible human carcinogen that same year. But they didn't alert the government to their conclusions in that regard. By 1993, the cancer data was even stronger. A second two-year rat study had confirmed that PFOA caused not only testicular tumors but liver and pancreatic tumors. In 1997, the very same year Earl was finally able to get EPA to start investigating the wildlife around Dry Run Creek, DuPont scientists coauthored a paper confirming that unless someone could prove that the precise way in which a chemical caused cancer in rats could not possibly cause cancer in humans in that same way, scientists could not discount the relevance of the cancer risk to humans.

It seemed that by the time Earl's cattle started dying, DuPont was roiling with internal conflict over PFOA. Their scientists and medical staff were alarmed and urging caution, management was holding firm to its insistence that there were no signs of a threat to human health, and the legal department was increasingly alarmed about the impending collision between these two incompatible stories.

As the picture took shape in my mind, I felt a lingering nausea. The evidence that PFOA was a health risk—and "risk" seemed an understatement—was overwhelming. Yet DuPont had kept pouring it into the environment for decades without warning the government or the people downstream. I thought about where we lived in Kentucky, just over the Ohio River from Cincinnati, two hundred miles or so downriver from Parkersburg. My boys, my wife, Sarah's mom, whose hair had just started to grow back after her round of chemo—we were the people downstream.

18

TEFLON PAWNS

Proving that DuPont knew that PFOA was toxic and that it was being released into drinking water still wouldn't be enough for our class action suit to prevail. I would also have to prove that all the nasty stuff they had been putting into the water was capable of causing actual disease among our class members, and I would have to make a jury *feel* the impact. As I wrestled with that issue, I kept coming back to this: two in seven. That statistic from the DuPont pregnancy study kept echoing in my head. Two babies born with eye defects in seven births. DuPont's attorneys and statisticians could argue that the sample size was too small to be statistically meaningful, but I had no doubt that to the women involved, it was more than meaningful; it was a lifelong nightmare. Through my discovery power, I forced DuPont to cough up the names of the PFOA-exposed working mothers in the birth-defects study.

They were Karen Robinson and Sue Bailey. Karen Robinson had been twenty-four and in her second trimester of pregnancy in 1978 when she worked just a few buildings away from the Teflon lab. Karen's job included crawling into an industrial dryer to clean out the debris. It was one of two identical machines that stretched the length of the production room. The dryers heated Teflon slurry as it shook across a screen, drying it into a fine white powder. To Karen, the powder residue she scrubbed from the dryer's inner walls seemed innocuous. It looked a lot like the laundry detergent she used to wash her clothes.

Karen had just celebrated her second wedding anniversary with her husband, the high school baseball coach, and this baby would be their first child. It would be a summer baby, a boy. After maternity leave, she'd have to go back to work, but maybe not back to DuPont. She had always dreamed of becoming a teacher. With a degree in education from Ohio State, she still hoped to follow that dream. She didn't love her industrial job, but she loved the people she worked with. And the benefits—health care insurance and maternity leave—made it even more appealing.

Karen had been a DuPonter for around a year by then. Like most new hires, she had paid her dues as a "utility man"—oiling railroad switches, shoveling snow, washing the plant manager's car. When a new batch of hires came in, she got to bid into a division. She chose filaments, where she packed up giant rolls of fishing line. After a while, she saw a posting for a Teflon opening, and she jumped on the opportunity. "Everybody said that's the best area to work," she said. "You don't have to pack up fishing line. You can pack out powder."

By the time she started working in Teflon in October 1977, she was pregnant. At first she was a "floater," bouncing around among tasks. Teflon products came in several forms, fine powder, granular, and liquid—all used for a bewildering variety of purposes and industrial processes, from molding plastics to lubricating bike parts to making carpets stain resistant to making dental floss glide through your teeth to making clothing and pizza boxes resistant to grease and stains. And yes, for coating frying pans. She rotated among them, depending on what needed to be done. When the lines shut down for cleaning, Zone 4 operators such as Karen had to crawl into the dryers and scrape them down. She also cleaned the claves— industrial pressure cookers that mixed ingredients into Teflon. They looked sort of like MRI machines, only bigger. "Time to clean the claves!" someone would announce, and she would slide onto a plywood board and scrape the wet goop into waste drums that would be shipped out for disposal. From time to time she had to hose down the trenches that filled with liquid dispersion. It sometimes splashed on her pants.

After a spell as a floater, Karen moved into a more permanent post in the fine-powder pack-out room. Teflon powder poured from the dryer chute into fifty-pound drums. Once they were filled, she sealed the drums, stacked them in the warehouse with a skid, and prepared them for shipment.

When she'd gained a little seniority, Karen graduated to mixing ingredients and operating the clave. Following the industrial recipe, she dipped measuring cups into containers of chemical ingredients, including PFOA, and poured them into the clave. Back then, the PFOA came in a powdered form, which felt like scooping sugar. The claves were up on the second floor behind heavy steel doors a few steps above the operating panel, where she controlled the buttons and gauges that operated them.

Something like magic occurred in the clave. Under the pressure and heat, electron bonds broke and new molecules snapped together in the chemical reaction that made Teflon. PFOA was a catalyst that sped up the reaction and stabilized the mixture. Without PFOA, DuPont believed, the Teflon production process would break down, making manufacturing more difficult, more expensive, and less reliable. When the polymerization was complete, Karen pressed a button dumping the batch into a coagulator that turned the raw dispersion into wet powder. The wet slurry was then heated and dried in the first-floor dryers (which, thankfully, she no longer had to clean). Once dry, the powder was ferried on a conveyor belt to the chute in the pack-out room. Occasionally a batch wouldn't gel quite properly, and it was dumped into a hole in the floor with a sump pump.

At the time, most of the workers who handled PFOA thought of it as being "like soap." But some of Karen's coworkers told her they'd heard that too much contact with the stuff might make you sick. Strict safety procedures that required measuring PFOA under a hood that sucked up any fumes made her think that maybe they were right. Sometimes the men on her shift offered to measure the PFOA, so she didn't have to touch it.

A few weeks before Karen's baby was due, upper management at the plant received an official notice about "Fluorosurfactants in Blood." It was dispatched to division superintendents with details on how and when to inform their workers. The notice began: "Through information provided by the 3M Company, DuPont has become aware that elevated organic fluorine levels have been detected in the blood of 3M workers exposed to certain fluorinated surfactants."

It went on to note that DuPont had purchased one of these chemicals, perfluorooctanoic acid ammonium salt—PFOA—for use in manufacturing Teflon.

"Our toxicological tests indicate that DuPont's fluorinated surfactants have a low order of toxicity. No known ill effects which could be attributed to these chemicals or C8 have been detected among employees in more than 20 years of experience with the products."

The notice declared that the company's handling procedures "have been designed to minimize exposure of employees to these fluorinated surfactants." And although "no known ill effects" had been associated with them, DuPont was reviewing their procedures, medical records, and toxicological information "as a precautionary measure."

The attached three-page Q and A answered eighteen questions that employees might ask, including whether Teflon-coated cookware presented a problem (no) and what the company meant by "low order of toxicity" ("a lethal [dose] would be about a cupful of eight fluid ounces"). And also this:

> *Will Du Pont be informing the appropriate regulatory agencies of the situation?*
> *At this point in time we see no significant risk associated with the fluorine content in the blood. The existence in blood has been known for 10 years and is published in open literature.*

In other words, no.

Division superintendents were instructed to first inform line supervisors in Teflon at 11:00 a.m. After lunch on June 27, 1978, at 2:00 p.m., wage-roll workers were to be told. That tier of employees included Karen Robinson.

Eighteen days later, on July 15, 1978, Karen became a mother. She named her son Charles but called him "Chip." He was lovely and, to Karen, almost perfect. His only defect had to do with one eye. His left eye had an extra fold of skin and a deformed tear duct.

. . .

In the spring of 1980, Sue Bailey worked one floor beneath the Teflon claves, next door to the fine-powder dryers. She sat in an office chair by a large, square hole in the floor. To her left were the trenches that Karen Robinson used to rinse out with a hose. Karen worked one floor above her now, cooking Teflon in the claves. The women crossed paths every now and then, but they didn't know each other well.

Sue was thirty-three years old, a mother of two, and the blue-eyed daughter of a veteran DuPonter. She was happy to follow in the path of her father, who had worked in Teflon for thirty years, ever since she had been three years old. He ran the extruders, pushing viscous polymer out in different shapes, like Play-Doh.

Growing up, Sue had seen her father suffer what the plant workers referred to as "the Teflon flu." But it had seemed no worse than the regular flu—fever, chills, aches, and nausea—and it eventually always went away. She knew the company gave him a little plastic box to protect his cigarettes, because a DuPont study in the early 1960s had found that Teflon-laced cigarettes could cause flulike symptoms. The plant didn't want any workdays lost to anyone getting sick or hurt. Once her father dropped something heavy on his foot, crushing his toe. He had been briefly unable to work. He had told Sue that the company had made him sit in medical for eight hours every day until he could do his job. That way, according to him, the perfect safety records stood because he was present at the plant.

Somehow, none of that discouraged Sue's dream of following her father in a DuPont career. Now she picked him up every day before work and they drove to the plant together.

Sue loved working for DuPont, not only because of the relatively handsome pay but because it was a door to a social tier above most other blue-collar jobs. "When you work for them, you have clout," she said.

Some people claimed you could walk into a bank and secure a loan just by saying you worked for DuPont. It wasn't just a company, it was a community, with a basketball team and group picnics to which everyone brought their families. The plant itself was like a little town, with its own fire department, medical center, and a cafeteria that served chicken cacciatore, one of Sue's favorites. The employees had locker rooms with showers, refrigerators, and microwaves. If she worked overtime, the company paid for a meal and a cab to drive her home. A gravy job.

Even though her dad had helped get her in the door, she'd had to work her way up like most everyone. She had started in 1978, in nylon, stringing gossamer-thin strands of fishing line on little reels overhead. Then she had bid into Lucite, where she had fed polymer into a machine that chopped it into beads. She'd worked all the shifts—day, afternoon, night. The plant never slept. She'd learned to sleep during the day.

While she was still a newbie in Lucite, Sue was "loaned" to Teflon. With no seniority, she had no say. But the company promised it was only temporary. Rumor had it that Teflon had been caught dumping something in the river. "If we get caught again," someone told her, "they're going to shut it down." It never occurred to her to wonder what was going into the river. She just knew she had a job to do, and she was going to do it well.

The Teflon job was easy. Sue was stationed in the dispersion pack-out area, a room filled with large metal tanks containing the liquid version of Teflon. She was usually the only one in the room, so she read a book to pass the time between batches. A few times a shift, the control room called down and said it was time to "dump a batch." Her job was to plug the pipe that emptied into the river.

Wearing regular clothes, she leaned into the big square hole in the ground, plugged the pipe with a stopper that looked like a rubber ball, and pumped it up with a bicycle pump. That formed a tight seal and blocked the drainpipe. She then turned on the sump pump, which channeled the stuff through another pipe that emptied into a pit on the plant grounds. She walked by it sometimes, a green slimy mess. She wondered if the pit was lined.

The only trouble was, sometimes the stuff overflowed from the cylinders and got all over the floor. That was a slipping hazard, against safety rules. The first time it happened, she called her supervisor and asked him what to do. He told her to "just squeegee it into the sump." But he warned her not to use water, because it would foam like a dishwasher with too much detergent. It was a mess, of course, and went everywhere, all over her clothes. She wore no protective equipment.

Sue was in Teflon for only a few months before being transferred back to Lucite. By then she was pregnant.

From the start, this pregnancy was not like her two previous pregnancies. The others had been easy. Now her hormones raged. Her insides were chaos. "I just had this gut feeling that something was not right." She was not by nature an anxious person. But she liked to feel in control. And now she felt very much out of control. She wasn't sick. She was worried. And she didn't know why.

As she entered the third trimester, she was overcome with a strange malaise. It wasn't a bug. It didn't feel like fatigue. She simply could not function. She didn't quite know how to describe it when she went to talk to her doctor. She just told him what her gut kept screaming: *Something is wrong!*

"I don't know what's going on," she told her doctor. "But if you don't get me some time off from work, I'm quitting my job."

"I don't know how I'm going to do that," he replied.

"I don't care," she said. "I cannot work."

He found a way. She took a leave. But she had to check in with the plant after every doctor visit. Thanksgiving passed, then Christmas. But the feeling never left her. If anything, it grew.

On a cold January morning, she felt the first contractions. Her husband went in to work, but she told him to be ready. The pain ebbed and flowed in terrible swells. At the hospital, she labored all day and all night. She felt as though her insides were ripping at the seams, but the doctors said it was too late for an epidural.

Then they gave her an epidural anyway, because they had to perform an emergency C-section. She was still awake when they lifted the baby into the chilled air of the delivery room. Everything was fuzzy when they laid her newborn son in her arms. At first she didn't see. But over the noise of the nurses and machines, she heard her doctor's words: "She's going to need a good pediatrician. Your baby has a birth deformity."

Sue snapped out of her postpartum fog. The newborn boy in her arms was blinking up at her with asymmetrical eyes. They were cornflower blue, just like hers. But those little blue eyes did not match. His right eye was misshapen and a little too low. Half of his nose was missing. The left side of his face was perfect. The right side was a Picasso. Sue studied her baby's face as tears streamed down her own.

"Don't get too attached," the doctor said. "He might not live through the night."

Sue cried as the doctor explained her son's birth defects. His eye was misplaced and malformed, with a "serrated eyelid." He had a "keyhole pupil" that resembled a tiny rip in his iris. The roof of his mouth rose into a peak. Was his brain okay? They didn't know. A swarm of white coats gathered around her son and stared, as if he were some specimen.

Holding her infant was terrifying. She had never felt so vulnerable, so afraid he would die in her arms. But she also felt a tidal swell of love. Though her vision was blurred by the tears that would not stop, the mother looked down and saw a sweet boy. A boy who, if he survived, would suffer trials she could only imagine. But a boy she would love fiercely. His birth certificate said William Harold Bailey III. But she whispered the name she would call him: "Bucky."

Within hours, Bucky was taken out of her arms and transported to Children's Hospital in Columbus, Ohio. The pediatric specialists there might know what to do. Heartbroken and healing from cesarean surgery, Sue was forced to stay behind. She shed more tears than she ever thought one woman could possibly hold. "I cried until there were no more tears to cry."

Ten days later, still sore from the scar on her belly, Sue walked with her husband into Children's Hospital to reclaim their son. They found Bucky propped up in a pumpkin seat. The doctors explained that he had trouble breathing while laying down.

"He just looked at me like 'Are you my mommy?' Of course it just melts your heart."

That didn't mean she wasn't terrified to take him home from the hospital. What if something happened? The doctors gave no diagnosis. "They had never seen anything like him." At home, he had a grand mal seizure, his little body growing straight and stiff as a board. It happened only once, but once was enough to create a lingering fear that it would one day happen again.

A week or two after Bucky was born, Sue's mother gave her a message from one of the DuPont plant physicians. A courtesy call, she figured. He was probably just going to congratulate her and see how the baby was doing.

When she called him back, she quickly realized he just wanted details about the deformity. But how on earth did he know? He said that any birth defects had to be reported by the company right away. She assumed he meant to the government. She told him about Bucky's eye and missing half nose. Bucky's eye defect and Sue's PFOA blood levels were among the notes DuPont's scientists reviewed as they collected data for Dr. Fayerweather's pregnancy outcome study in 1981—the study that was never finalized or reported.

. . .

While Sue was on maternity leave, Karen Robinson was pulled out of her job in Teflon. The "employee communications" memo she was handed gave a lengthy explanation. It led with news of the PFOA study of pregnant rats and concluded:

> *Women of childbearing capability will be allowed to bid for other plant jobs after a permanent plant posting has been made.*

Karen was one of about fifty women at the plant who fell into that category. Her exposure was not hypothetical. She measured PFOA powder and poured it into the Teflon claves. She had crawled around in the fine-powder dryers, cleaning them, inhaling the particulate. She had hosed dispersion out of troughs and scraped it from the inside of the claves. And she had done those jobs during the second and third trimesters of her pregnancy with Chip.

"It could affect females," she was told, somewhat belatedly. "We're going to move you out as a precaution."

The announcement stirred up great concern. So did the immediate blood testing of the women. That's how Karen learned she had PFOA in her blood—2,500 ppb. She was worried. All the women were. What did her blood level mean? Was it high? She was never told, but DuPont's scientists had actually recommended in April 1981 that anyone with more than 400 ppb in his or her blood be removed from the workplace. Karen's blood levels were *six times* that level. Her worry intensified when she read the list of thirty-nine questions and answers included in the announcement. One of them in particular alarmed her.

What were the birth defects noted by 3M in the unborn fetus?
Eye defects are reported but complete testing will be required.

She thought of Chip's eye.

She met with Dr. Power and told him about Chip's eye defect. Her son was two and a half years old now and otherwise healthy. But if the baby rats had had eye defects, could that have something to do with Chip's eye? Together, Karen and Dr. Power drew a sketch of her son's eye defect and put it into her file, where the results came to light again in 1981 as part of the Fayerweather study.

The two babies out of seven that DuPont had found with eye defects in 1981 had names: Bucky and Chip.

19

ACTUAL MALICE

Summer 2002
Cincinnati, Ohio

S tories like Sue's and Karen's would move a jury, but I wanted more evidence to show that DuPont had been more than just careless in their handling of PFOA. For that, I wanted something showing an actual awareness of the problem on the part of the company and a conscious choice to ignore it. The document I now held in my hands, a summary of a meeting that had occurred eighteen years earlier, appeared to be just that. I could scarcely believe what I was reading. After a year of digging into an unyielding mountain of documents, I felt as if the earth had given way beneath my pickax and sunlight had flooded into a long-hidden cavern filled with glittering treasure. This changed everything.

The document told this story: On a Tuesday in late May 1984, top DuPont execs had held a closed-door meeting to decide how to respond to a critical development in the division that made Teflon. The meeting had followed the circulation of memos from corporate medical director Bruce Karrh, who had been concerned for at least two years about a "great potential for current or future exposure of the local community" from PFOA emissions leaving the plant. The meeting had been convened because his concerns had been proven justified. Washington Works had dispatched a handful of employees off the plant to discreetly gather water samples from different spigots outside the plant, around the community. Armed with plastic jugs, they had been instructed to fill them from taps in gas stations, local markets, and the like. They had gone seven and a half miles upriver from the plant and seventy-nine miles downriver.

Of eleven samples analyzed by DuPont's in-house lab for PFOA using gas chromatography, the two taken from Washington, West Virginia (which used Lubeck Water), and Little Hocking, Ohio, came back positive. Now it was no longer just Karrh's anxious speculation, it was a fact: DuPont had contaminated the water of the plants' neighbors.

It was an existential crisis. In 1984, Washington Works was the largest DuPont plastics plant in the world, and Teflon was one of its leading products, growing in volume and profit every year (and nowhere near its peak). PFOA was considered an invaluable chemical in the manufacturing process of Teflon; it had been used—and emitted—at Washington Works for thirty-two years, and also used at Parkersburg's two sister plastics plants in the Netherlands and Japan. Warning signs about PFOA's toxicity had been stacking up for the past quarter century. DuPont had known for twenty-four years that PFOA showed signs of toxicity in rats exposed to it. Subsequent studies had shown similar effects in dogs and monkeys. The company had known for five years now that PFOA was persisting and building up in workers' blood. They had noted eye defects in the children of two of the seven women who had given birth after exposure in the Teflon division. Now that they knew PFOA was contaminating the local community's water supply, what were they going to do about it?

At that Tuesday meeting in DuPont's horseshoe-shaped high-rise in Wilmington, nine executives from the fluoroproducts business unit and Gerry Kennedy, DuPont's lead in-house toxicologist on PFOA issues, debated the options. But first they reviewed the safety provisions that were in place to guard against on-plant exposure. Measures to protect workers through "engineering controls and protective equipment" had been implemented since 1980. In one respect those on-plant safety measures seemed to be working: the PFOA levels in workers' blood were decreasing. The execs were briefed on additional engineering changes to further reduce worker exposure. The fine-powder dryers could be adapted with an exhaust system that would capture a stream of PFOA-rich air and funnel it to the exhaust stacks. "The intent is to first reduce in-plant exposure, and second leave a future capability for treatment of this relatively concentrated stream." For the time being, at least, it seemed that the company was simply redirecting PFOA pollution away from the plant and sending it into the community. And with the anticipated increased production of Teflon, the PFOA pollution caused by waste from the production process would presumably only get worse.

The fact that the company now knew PFOA had made its way into the community's drinking water had raised the matter to a crisis. The individuals gathered in the room noted strong differences of opinion among the different divisions of the company on what the response to the contamination should be.

Those in the meeting believed that the medical and legal departments supported the complete elimination of PFOA, fearing the chemical's potential danger and the huge liability it could create, with the business leaders being more inclined toward the opposite point of view, concerned that a less toxic replacement chemical was still speculative and that by eliminating PFOA they would "essentially put the long term viability of the bussiness [sic] segment on the line."

I blinked hard and looked again to see if I'd read that right.

It appeared that business and science were at odds. Whoever had drafted the memo (1) must have flunked spelling in grammar school, and (2) anticipated that the struggle between the company's scientists and lawyers against its bean counters would only intensify.

Though the executives concluded that it was "not an easy and obvious dicision [sic]," there was a consensus reached that "the issue which will decide future action is one of corporate image, and corporate liability." As for this liability problem, it was agreed that "liability was further defined as the incremental liability from this point on if we do nothing as we are already liable for the past 32 years of operation." In other words, why forgo a hugely profitable business segment, they seemed to be asking, if you were already on the hook for thirty-two years of pollution? And who ultimately prevailed in this debate? Given that DuPont continued to not only keep using PFOA but actually increased its use and emissions—dramatically—over the next decade, it seemed clear that the corporate bean counters and business leaders had prevailed over the company's own scientists and lawyers.

. . .

I felt the case I had already built showing the dangers of PFOA and DuPont's negligence in ignoring the problem was formidable—so formidable that I was confident our medical-monitoring claim would prevail. For that, simple negligence (tortious conduct) was the key: DuPont had a duty not to pollute the water and had failed to act as a reasonable corporation would. To prevail in that regard, I didn't need to prove a specific intent to do wrong or conscious awareness they had been doing so, only that the company had done something wrong and thus had breached their basic standard of care to the community.

DuPont's attorneys' counterargument went like this: We didn't breach our duty because we had no reason to believe there was any problem. Therefore, we didn't act unreasonably. Basically, there is no tortious (wrongful) conduct because we had no clue that there was any issue with PFOA.

But here in my hand was a document from corporate files that neatly and succinctly summarized in DuPont's own words that they had known about the PFOA problem and debated what to do about it, worried about how much it would cost, worried about undermining the economic viability of their Teflon division. Here was a window flung wide open showing the soul of a corporation at a moment of truth, as they contemplated decisions that would lead to the situation in which we found ourselves today. Their subsequent actions, which I could now fully document, showed that even knowing of this problem, they had ultimately chosen not to eliminate PFOA and hadn't taken all the remedial steps discussed in the meeting. They hadn't scrubbed all the PFOA out of the stacks. They hadn't switched all lines over to a different, safer chemical. They hadn't reported their concerns to the government. Two decades later, they were *still* pumping PFOA into the environment.

Not only did this neatly dispatch the company's specious claim that they had no reason to suspect that PFOA posed any risk of harm, it opened the door to another, much more serious level of legal liability. Now if we went to trial, we could claim not only negligence but conscious disregard, which in legal language often has the appropriately scary designation of "actual malice." To prove actual malice, we would need to show, in essence, that the company had been aware of the risks and acted with knowledge of the risks—exactly what I thought the memorandum of that meeting depicted in startling detail. Now, if we won our case, it would open the door for us to seek punitive damages on top of any medical-monitoring or other relief we might win for the community. As someone who had represented chemical companies, I knew there was little that companies feared more than to be facing punitive damages. Punitives, as they were abbreviated, opened the door to a massive level of liability. Unlike the "actual damages" the Tennants had had to assess in their case, punitive damage awards could greatly exceed—by several multiples—the value of any actual damage caused. If a jury were to find conscious disregard, the judge could then convene a whole new phase of the trial, the sole purpose of which would be to determine just how much money the company had so the jury could award a large enough additional amount of damages to "punish" the company sufficiently to ensure that it would never consider behaving so unconscionably again.

This was the justice that Earl had longed to see, accountability for unconscionable behavior. Punitives weren't just about the dollars; they were about holding a company responsible for consciously doing wrong.

I felt light-headed; without realizing it I'd been so stunned by the contents of the document that I had been practically holding my breath the whole time I'd been reading and rereading it.

The memo validated everything I'd long believed to be true and had been working for nearly six years to prove, that PFOA might pose a risk of serious harm and that DuPont had known it. They had known it when PFOA contamination was killing Earl's cows, even as they tried to blame the deaths on Earl. DuPont had known eighteen years ago that this "forever chemical" was contaminating public water and had known the risk it might pose to the public.

20

HAIL MARY

Fall 2002
Cincinnati, Ohio

I t had taken more than a year, multiple court-imposed deadlines—most of them missed—and my constant pressure for more documents until a threat of sanctions, and my small band of co-counsel, had finally gotten us to this point. It reminded me of the old adage "Be careful what you wish for—you just might get it." Reviewing the new documents would be a massive undertaking—450,000 pages of new material—a cache so vast that DuPont had enlisted no fewer than fifty-six attorneys just to review them and produce them to us.

In DuPont's rush to meet the court's deadline, some documents must have slipped through the cracks. I realized that when the company began asking for things back. Some privileged documents, they claimed, had been produced by mistake. Some of them included emails or memos among DuPont's attorneys or to their client discussing the PFOA matter. DuPont wanted them back. Immediately.

During the course of the litigation, we had usually resolved such issues amicably. But this time, DuPont's lawyers had steam coming out of their ears.

I told DuPont we would not be returning some of the documents because they had waived privilege over them—either expressly in letters to us (after we'd questioned why they were privileged in the first place) or by producing similar documents on the same subject. Not only that, I demanded that DuPont turn over the rest of their documents on those same subjects—from both this case and the Tennants' case. A whole new round of fireworks began. This was now an issue Judge Hill would have to decide.

I can only imagine the holy hell that broke loose behind closed doors somewhere deep inside DuPont over the waiver issues. Some of the documents the company had turned over from the lawyers provided incredibly vivid and clear insights into the internal conflict within DuPont and the increasingly fierce struggle among the legal, medical, and business departments.

I'd have a long wait before I found out if I could make use of the disputed documents as evidence. They were enlightening in any case but not of great practical value if I couldn't present them or refer to them in court. But even the undisputed documents provided plenty of revelations. One thing in particular became clear: several of the Spilman lawyers who had represented DuPont were now regulatory officials at the West Virginia Department of Environmental Protection. After working on PFOA issues, they had left the law firm to accept their new government posts. Spilman had been the liaison between DuPont and the state during negotiations on the latest state consent order that had created the whole CAT Team process. (We now called it the CAT Sham.) In fact, the Spilman lawyers had helped draft the very consent order their new employer was now enforcing.

I had heard Ed and Larry joke about the revolving door between West Virginia government and industry. But government officials usually left regulatory agencies to take better-paying private sector jobs. Apparently, the door was now swinging in the opposite direction. People steeped in the corporate culture and perspective of DuPont might not be the best candidates for enforcing environmental regulations against their old friends and colleagues.

With my new understanding of the incestuous relationship between

DuPont and the West Virginia DEP, I urged the agency to enlist an independent organization to oversee the still ongoing consent-order work between DuPont and the state. It ignored my request. I also reminded the agency of its promise of an independent investigation into its document-destruction practices. Nothing came of that, either. Not that I really expected it to. By this point, I had concluded that much of my correspondence with the state government was going into its circular files. As Harry, Ed, and Larry reminded me: that was how things worked in West Virginia.

I was well beyond frustrated—and not just about the revolving door. DuPont had struck a serious blow in the CAT Team skirmish with their 150 ppb safety level. Since that absurdly high level had been announced, the state agencies in West Virginia had seemed perfectly content to use it as an excuse to do nothing further. *No problem here!* I sent new letters to the feds, pleading with them to reject the CAT Team number and take action.

Silence.

Meanwhile, I kept attaching documents about PFOA toxicity and environmental proliferation to our public court briefs and to my letters sent into the public docket EPA had created. Even as I was building what I felt was an increasingly compelling case for the class action, I felt I owed it to the people I represented to continue trying to push the government regulators to force DuPont to take action to stop the ongoing public health threat and at least provide them access to clean water immediately. As the case dragged on—the judge would soon set a trial date for July of the following year—I felt as if I were in the same place Earl had been when his cattle herd was dying. Every day that passed without action was another day of excessive community exposure to the chemical. It ate at me. I wondered if the regulators dragging their feet had children of their own drinking PFOA. And here I was with my files bulging with powerful evidence—the 450,000 documents that painted such an alarming picture of DuPont's actions and their effect on thousands of people—just like Earl with his cardboard box full of evidence that nobody would look at.

DuPont's countermoves had been effective; I was losing ground, and

I knew it. By putting up the smoke screen of 150 ppb, outrageous as that might be, they had reclaimed the upper hand, and the momentum was now with them. I needed to do something to turn things back around. Someone needed to act, especially if the regulators weren't doing anything. As I considered all that I could prove now, based on what I had already uncovered, a bold plan occurred to me. I had the evidence I needed to prove our medical-monitoring claim under West Virginia law—significant exposure, hazardous substance, significantly increased risk of disease, wrongful behavior by DuPont—using nothing but the company's own documents. Why should I wait around for trial when we could take action now, in court, to fix this mess? By relying on DuPont's internal statements, studies, and admissions—facts they couldn't plausibly deny—I would have a shot at short-circuiting the whole trial process by going straight to the judge to seek summary judgment.

In a complex case like this, asking for summary judgment would be an extraordinary move. But the more I thought about it, the more I realized I just might be able to pull it off. A lawyer moves for summary judgment when he or she is confident that there is no genuine dispute about any material facts. Facts can almost always be disputed, even if the basis of dispute is weak. No "genuine dispute" on any "material fact" is a very high bar to clear, which is why summary judgments are rare, especially in hugely complicated, fact-filled cases.

But those documents. The beauty of my plan was that I wouldn't need to trot out expert witnesses or reports from hired guns to make any of the required points, all points DuPont could dispute with their own hired guns. They could pick apart outside experts, but how could they dispute their own internal documents? Over the years, DuPont themselves had established all the necessary legal points. Significant exposure? Well, sure, that's what all the water and blood testing they'd ordered had been about. A hazardous substance? Twenty-five years' worth of animal studies and an internal classification as a confirmed animal carcinogen and possible human carcinogen should fill that bill. Exposure significantly increased the risk of human disease? DuPont had been doing worker studies for years to delve into that issue. Would a doctor recommend medical testing for the disease? That was exactly what DuPont's doctors had been doing for their workers since the 1970s.

What I wanted to get across to Judge Hill was quite simple: if the presence of PFOA in DuPont workers' blood caused enough concern to warrant ongoing medical testing and studies by DuPont's own scientists, then community members exposed to the same stuff—at levels much higher than DuPont's internal guideline—deserved the same precaution. How could DuPont dispute the findings of their own scientists?

I filed my motion in March 2003, and the judge set an April 18 hearing date. I spent days poring over my files, picking documents that made the point. I felt my case was strong, and even if I failed, I believed it would highlight for the judge the ongoing harm that was being done. PFOA was still flowing into the water, still streaming into the air, still accumulating in the bodies of my clients. Judges have enormous power; if I made it clear enough that the situation was not only ongoing but getting worse, he might be moved to do something, even if it was short of summary judgment.

Before the hearing, I was going to need a wardrobe overhaul. For years, Sarah had teased me about my shoes. I don't care about clothes, and I care even less about shoes. I had two suits, a collection of white button-down dress shirts, some ties, and a pair of battered black wingtips. Sarah used to warn me that I'd never become a partner at Taft if I wore those shoes. I'm guessing those shoes didn't make the difference, since I'd been made partner a few years back. But there was no need to press my luck. This was an important moment. It couldn't hurt to have new duds, so I got a new suit and shoes.

Once again I made the familiar drive to Parkersburg to the Wood County Courthouse, a 1950s concrete-slab architectural atrocity two blocks from the Ohio River. As I walked up the concrete walk to the entrance, I cast a look in the direction of the river, imagining the PFOA flowing downstream in the powerful current.

These hearings are nothing like the crisp, dramatic spectacles often portrayed in the movies. There is no jury to impress. The opposing lawyers sit on opposite sides of the courtroom and studiously avoid making eye contact. The dialogue back and forth with the judge is extremely technical and often expressed without emotion. Instead of shouting or gesturing, people politely clear their throats. But internally, in the minds of the opposing counsels, the electricity is intense. I have no doubt that DuPont's lawyers—led again by the regal Larry Janssen—were surprised and outraged by my summary judgment motion, and they intended to stomp on it until it was dust under their feet. They were also still under threat from my request that the judge issue sanctions against them—both monetary sanctions, for the considerable time we'd spent trying to force them to turn over discovery, and legal sanctions, in the form of telling a future jury that they could assume that the documents destroyed by Kennedy, DuPont's toxicity expert, which no jury would ever get to see, had proved that PFOA was a hazardous substance. I was happy to be keeping the heat on.

I was up first and immediately began laying out the reasons for considering PFOA to be toxic. If I'd had any illusions that this was going to be a slam dunk, Judge Hill instantly relieved me of them. "Why do you not have to show that at least someone has been adversely affected by C8?"

"Your Honor—" I began to answer, but he shut me down with a flat statement, which, if I couldn't rebut it, would stop my motion right then and there.

"It is not just a given that C8 is a hazardous substance."

"We believe it is."

"How is that?"

I launched into it: all the supporting documents I had filed with my motion.

As I went through the many items indicating DuPont's concerns over whether PFOA was making their people sick, Judge Hill stopped me with questions about one study that had showed unusual liver-test results among plant workers after exposure to PFOA and concerns that those results might be a precursor to liver disease.

"Wait," he said. "Is that sufficient to grant relief, just the fact that it may? It doesn't have to be probable, at least probable?"

That was a key moment. The judge was focusing on the part of the West Virginia Supreme Court of Appeals' decision recognizing the viability of a medical-monitoring claim. I needed to help the judge see why medical monitoring was different from the usual claims he heard. In this case, the relief I was asking for wasn't to help someone who had already been diagnosed with a disease from exposure to a chemical, it was to help people learn as early as possible if they were at increased risk of getting sick due to exposure. Once it was generally agreed by scientists that a chemical was capable of posing a significant risk of serious disease (and I felt that DuPont's own records established that fact), that was enough to grant medical monitoring. I didn't need to prove that the chemical had already caused disease in any person in the past or that it was causing it now.

"Medical monitoring is not to recover for the actual damage," I told him. "It is: 'Is there a significant increased risk that can occur?'"

But as I knew they would, DuPont's team countered with the atrocious CAT Team assessment of a "safe" level of PFOA exposure in drinking water of 150 ppb. Janssen's co-counsel Steve Fennell took over to drive that point home.

"West Virginia, it put ten toxicologists on it in May of 2002, and that is medical and state toxicology, and they determined after reviewing all the evidence that 150 parts per billion was the safe human health level. The amounts we are talking about that are being found locally are down in the one or two parts per billion level." In other words, even if PFOA could cause some disease, the levels in the water locally were so low, according to the revered CAT Team, that they could not possibly present any risk of ever causing disease.

Of course, he neglected to mention that the members of the CAT Team had included DuPont's own scientists, but there was no need to point that out because the judge brought him up short.

"I think the trouble with this argument among lawyers about all of these issues, they are not technicians or scientists and they don't know. I don't have any idea if what they're telling me is true, or if they even know what is true. That is why in this particular case it seems like a summary judgment would be . . ." He paused, and I held my breath. Just as it seemed that he was about to say my motion was denied, he swerved in another, much more favorable direction. "Except for the fact that the plant is continuing to discharge this dangerous, or potentially dangerous, chemical, it is a case where summary judgment would probably be rejected and we'd be back where we were, and I'm not so sure the jury wouldn't come down harder on DuPont than I would."

My brain was in hyperdrive. Time seemed to slow. I was on the cusp of defeat, yet with a surprising potential for victory. Clearly Judge Hill didn't think that the facts, at least with respect to the necessary element of increased risk of harm, were indisputable, as required for summary judgment. Even if it was DuPont's own internal studies versus the CAT Team findings, there were still two rival sets of facts in evidence, no matter how tainted the latter was. So I knew that my chance for summary judgment was gone. At the same time, I had hit home with my secondary strategy of making the judge see the unremitting, ongoing, real-time potential harm to his neighbors. I could see he was sympathetic to my clients' predicament. I had to come up with something, and fast, short of summary judgment that would still allow him to grant some relief. Then it came to me, a wild shot from the hip. If I couldn't get summary judgment because the facts could still be disputed, maybe I could instead ask the judge to issue an injunction ordering DuPont to do something that would address the ongoing harm.

"If Your Honor is concerned about the standard for summary judgment," I said, "we submit that it would be appropriate for injunctive relief."

"That is what I've been thinking about," Judge Hill said. The magic words.

Janssen stood back up. Sensing that things were going south, he had to head the judge off before he said something final.

"Your Honor, can I go back to basics for a moment?"

For him, the basics were the bottom line of all DuPont's defense: no evidence of harm to human health. If the state of West Virginia had determined that it took at least 150 ppb PFOA in the water to be unsafe, our class's exposure to PFOA—none of whom had PFOA levels anywhere near 150 ppb in their water—was not "significant," as required by medical monitoring law.

I interjected. The issue wasn't just PFOA in the water. It was PFOA in my clients' *blood*. I pointed to an internal model DuPont scientists had created in 2001 to estimate the level of PFOA in a person's blood based on the amount of the chemical present in the air or water. That model predicted blood levels in the hundreds or even thousands of parts per billion for those living around Washington Works, astronomically higher than the approximately 4 to 5 ppb that 3M had reported finding in the general population's blood back in 1999.

Janssen countered that we could not use DuPont's model to satisfy our burden of proof. That had been nothing more than an estimate, he said. If we were going to argue "significant exposure" based on blood levels (and not water numbers), we needed proof that our class members had significantly elevated blood levels.

The problem was, I knew of only one lab in the country that had the capability to test blood for PFOA. That lab was under exclusive contract with DuPont and would not run tests for us without DuPont's permission, which they would never give voluntarily.

Without a lab, we had no way of getting the data DuPont was now saying we needed to make our case. A catch-22.

"This court has the authority through injunctive powers to order this [blood] testing," I reiterated.

Janssen's eyes lit with fire. "I don't agree with that," he said.

Judge Hill narrowed his gaze, equal parts disapproval and bemusement. What was it Janssen didn't agree with? he wanted to know. "You don't think I have the power to do that?"

Janssen knew he was in a slippery situation. He gently eased back, just a bit. "I would really have to think about that," he said. "I hate to tell any judge that he doesn't have the power—"

"You wouldn't tell a federal judge that, would you?"

This was clearly not going Janssen's way, and he had managed to irritate the judge to boot. Judge Hill certainly wasn't intimidated.

"I could order this testing to begin," Hill suggested, "and then you could have an expedited appeal to the Supreme Court of Appeals to see if that judgment was right. Meanwhile, this stuff is being spewed out into the water and air."

So that was it. I didn't get summary judgment, but in winning the motion for injunctive relief, I had succeeded in changing the momentum of the case back in our favor. Now DuPont was once again on the defensive. And the injunction ordering blood testing for the people being exposed—at least pending appeal—would provide one kind of relief to my clients, relief they had craved ever since learning that their water had been contaminated. They would get to know at last if their bodies had been compromised, and if so, how badly. The testing would also provide the essential blood data we now knew DuPont was going to demand we produce if we refused to accept the CAT Team's 150 ppb water number as the test for "significant exposure." I had also gotten some very important advance intel on how DuPont was planning to attack our case at trial. If I hadn't taken that shot and pushed for summary judgment now, I would likely not have known of DuPont's plan to insist on the need for actual blood data before it was too late to collect it.

Before the hearing ended, Judge Hill addressed several other pending issues, including our complaints over the way DuPont had been stalling handing over documents requested in discovery. He granted my request for sanctions. As "punishment" for Kennedy's earlier destruction of evidence in the CAT Sham, he ordered DuPont to pay our lawyers' fees for the time we had spent fighting those issues. It wasn't a lot of money, but it was a symbolic victory. More important, he ruled that DuPont would face a "negative jury inference" at trial. The jury would be instructed to assume that whatever documents had been destroyed had held information incriminating to DuPont.

These were two major wins for us.

Well, one and a half. Before the first drop of blood was drawn for testing, DuPont took the judge up on his invitation and appealed the injunction. The Supreme Court of Appeals of West Virginia would have to decide. The blood testing was put on hold.

21

NARRATIVE WARFARE

Spring 2003
Nationwide

With the blood testing on hold and my impatience with the glacial pace of the judicial process and the regulators going through the roof, I decided I needed to do something on behalf of my clients, something beyond the legal case. Going for summary judgment had been a power move, and it had paid off. The momentum of the case was clearly back on our side, but what practical good was that doing for my clients right now? The thousands of people I had promised to help were growing increasingly impatient with the slow grind of litigation. They were peppering my West Virginia counsel with irritated calls and, what was worse, wondering if every random pain or stomach upset was the product of the water many of them were still drinking. They weren't wealthy people, and the expense of bottled water could still trump the unknown risk of future illness. I found it intolerable that they had to make such a choice and that regulators were not forcing DuPont to pay for bottled water for them while the regulatory investigations and legal case were under way.

I was frustrated that EPA was not trying to help these people, not even by challenging the bogus CAT Team number. If the people at the agency had read my letters, all those carefully organized exhibits and attachments laying it all out for them, they would certainly know just how ludicrous a 150 ppb guideline was.

As my exasperation grew, I thought of Teddy, Charlie, and Tony, now five, three and a half, and almost two. Teddy would be starting kindergarten in a few short months. Beyond my own family, I was now connected with the Tennants, the Kigers, and their neighbors in a way I had never imagined. I couldn't stop fighting for them now no matter what it took. As I consulted with Sarah one evening, curled up on the couch over a bowl of buttered popcorn, we both decided that this was what I was meant to do.

I refused to give up. Inaction meant simply that I had to bring even more pressure to bear on EPA. Finally, in September 2002, my persistence paid off and the agency began to react. My letters, loaded with all the water-testing and toxicology data I'd dug up, had raised enough questions that EPA now knew it could not just sit back and accept the CAT Team's work without independently verifying it. It very quietly launched its own internal investigation, a "priority review" of the toxicity of PFOA, and created a new public docket to serve as a repository for PFOA risk and toxicity information.

I needed to capitalize on this foothold. First I doubled down on my efforts to push EPA to continue to study the issue independently of DuPont. I bombarded it with more letters highlighting the relevant points I had worked so hard to grasp. To each letter I attached additional supporting documents, most of which had come out of DuPont's internal files. Into the public docket they went. In spite of all I had been through, I was still infected with Earl's stubborn faith. Regulators just needed to see *all* the data, I told myself. If only they would look at the complete data set, they'd see what I was seeing, and they would no longer be able to ignore the obvious danger. Then they would understand that the CAT Team had been wrong.

I considered my barrage of posts to EPA quite literally a public service. I believed I was doing EPA a favor by doing the legwork and sharing the fruits of my considerable (and very expensive) labor. I was saving some lucky government employee thousands of hours of obtaining, reviewing, and organizing the data. And that's if he or she could even find it. Much of what I was sharing was hard-won internal DuPont data obtained only through all my discovery battles and would not turn up in a search of publicly available documents. I'm certain that a few EPA employees were grateful for my contributions. But not everyone felt that way. Years later an agency source would tell me in confidence that my earnest efforts had outraged certain of their colleagues. To them I was the annoying gadfly who had burdened them by dumping work onto their already overfilled plates.

To hell with that.

In addition to pressuring EPA into action, my letter-writing campaign, my steady uploading of documents to the EPA public docket, and my court filings had another important objective: I was making sure that key insider documents—nonconfidential unpublished studies, internal memos, meeting notes—were available not only to the court and regulators but also to the general public. Getting word out to the public about this public health threat was key, and I was counting on my documents' grabbing the attention of interested parties and organizations that possessed a bigger bullhorn than I did.

It worked. The organization that stepped up and would assume that role for me was the Environmental Working Group (EWG). Based in Washington, DC, the EWG is a nongovernmental organization run by scientists, policy experts, lawyers, and communications experts who gather and vet the science on issues related to human health and the environment. Its people are expert at aggregating data and translating the science into terms nonscientists can understand. The group bills itself as nonpartisan, though some critics (especially the chemical industry) disagree.

As part of a 2003 series called "Pollution in People," the EWG zeroed in on perfluorinated chemicals (PFCs), the family of chemicals that includes PFOA and PFOS.

It had learned about our class action case—and our court filings. For its report, the EWG said, it had reviewed fifty thousand pages of documents. Some had been obtained from EPA. Others were part of the "growing body of independent scientific studies." And the bulk of them were "internal documents from DuPont and 3M disclosed in ongoing litigation."

That would be my litigation. All the documents I had submitted—attached to court filings and to my letters to EPA—were now (with some exceptions) in the public record. And now that they were in the public record, I was free to arm the EWG with all of it.

On April 3, 2003, the EWG posted a story on its website with this headline: "Toxic to Animals and People. Persistent Forever. Pervasive in Human Blood. One Perfluorochemical Has Been Banned. Another Is Under Regulatory Pressure. It's Time to Take a Closer Look at . . . PFCs: A Family of Chemicals That Contaminate the Planet."

It was immeasurably gratifying to have an organization outside Du-Pont (and the government) make use of the materials I had shared, scrutinize the data—and see what I had been seeing. That science was starting to speak for itself. And it was saying things that didn't jibe with DuPont's ongoing message. "No evidence of adverse health effects" was not going to fly in the face of increasingly prolific evidence to the contrary.

The EWG pulled no punches in spelling out the stakes. "As more studies pour in, PFCs seem destined to supplant DDT, PCBs, dioxin and other chemicals as the most notorious, global chemical contaminants ever produced," the report stated. "Government scientists are especially concerned because unlike any other toxic chemicals, the most pervasive and toxic members of the PFC family never degrade in the environment.

"Every new molecule of PFOA produced by the chemical industry in the coming years will be with us forever," the report said. "Non-stick pans, furniture, cosmetics, household cleaners, clothing, and packaged food containers contain PFCs (perfluorinated chemicals), many of which break down into PFOA in the environment and in the human body."

Within days, the story flooded the national news. PFOA had officially mushroomed from a "local problem" into a matter of national concern.

For a brief, exhilarating moment, I felt relief. I was no longer the lone "crazy plaintiffs' lawyer" flying in the face of the "experts." I was not misreading the internal records. I was not delusional.

The sudden flood of news alerts about PFOA didn't just serve to make me feel optimism for the first time in a long—very long—time. It seemed to push EPA over the edge to finally do what three years of unceasing effort on my own, without the supporting voice of an independent outside organization, had failed to accomplish.

Just eleven days after the EWG report, EPA announced on its website: "The Agency will be conducting its most extensive scientific assessment ever undertaken on [PFOA]." The process would "guarantee that any future regulatory action on PFOA is protective of public health and supported by the best scientific information." Included in the announcement was the revelation that EPA had prepared a draft risk assessment calculating "margins of exposure" suggesting that PFOA might present risks to human health, particularly among young women and girls.

EPA's announcement launched a new round of splashy nationwide headlines: "EPA Steps Up Study of Teflon Chemical Risk to Humans" (Reuters); "Teflon Is among Products Under Scrutiny for Health Risks" (*Wall Street Journal*); "Teflon Coming Under Fire" (Associated Press).

This time the target wasn't some alphabet-soup chemical nobody had ever heard of. It was a product nearly every American had in his or her kitchen: Teflon.

I could almost feel the shock waves reverberating in Wilmington. I assumed that every headline that mentioned "Teflon" and "health risks" fell on DuPont headquarters like a mortar. DuPont officials would not discuss Teflon profits, but in my November 2002 deposition of Richard Angiullo, a DuPont vice president and general manager of fluoroproducts, he had revealed that products made using PFOA had generated $200 million in after-tax profits in the year 2000 alone, roughly 20 percent of the company's annual net profit. That kind of money would be hard to replace.

Even the trade journals were blasting DuPont. A *Plastics News* editorial criticized DuPont for failing to disclose the drinking water contamination, calling the Parkersburg situation "a lesson on how not to build community trust."

This was no longer a local tiff in some backwater state court. It wasn't just one delusional lawyer throwing stones at an armored giant. It was no longer a matter that could be contained by state regulators with questionable ties to DuPont.

Federal regulators were closing in. The media smelled blood and were swarming. DuPont executives were likely huddled around a table somewhere, trying to hold down panic. This was a break-glass-in-case-of-emergency situation. I wasn't sure what DuPont's next move would be, but I was certain they weren't about to give up.

. . .

Under storm clouds of bad press, DuPont launched a PR counteroffensive—and even that was a gamble. Though the constant refrain "There are no adverse human health effects" may work in the courtroom, it isn't exactly the catchy jingle you want to trumpet to consumers. DuPont didn't want to raise concerns about human health and PFOA where they did not yet exist. But that was a necessary risk to take to get out in front of the issue, set the tone, and shape the debate. Waiting around to defend the next round of bad press could be disastrous. To that end, DuPont developed a plan to offer one-on-one briefings with key reporters at the powerful media gatekeepers: the *New York Times*, the *Wall Street Journal*, newswires, and press syndicates.

DuPont immediately and directly hit back at the EWG—and EPA—in a press release. They said that EPA's allegations about risks were "a clear misinterpretation of the data."

The release claimed that there was "extensive scientific data" to support their claim of "no evidence or data that demonstrates PFOA causes adverse human health effects," including "worker surveillance data, peer-reviewed toxicology and epidemiology studies, and expert panel reports." Further, there were "many studies on the toxicology of PFOA leading us and others to conclude that the compound is safe for all segments of the population, including women of child-bearing age and young girls."

I always took note when DuPont referenced the "many studies" and "extensive data" supporting their "no evidence" claim. If these studies did in fact exist, why hadn't they sent me any in discovery? Every time DuPont mentioned a study, I requested a copy. When the alleged exculpatory study would arrive, if I skipped to the conclusion, it would almost always be the same: no evidence of health risk. Later, when I asked my experts to evaluate the data that preceded the conclusion, they would often say: This data does not support the conclusion.

Despite the growing number of negative stories in the media, it appeared that DuPont still had some of the press in their back pocket. The local TV station in Parkersburg jumped into the fray to defend DuPont (and take a swipe at plaintiffs and their attorneys) with an editorial by the station manager:

> *Despite the claims by some folks more bent on getting rich than getting to the truth, there has been no evidence produced that shows that C8 is "harmful, deadly, or poisonous" to humans.*
>
> *If it were proven that DuPont had knowingly withheld information about the hazards of C8 to humans . . . or showed a callous disregard for C8's impact on people and the environment, I'd be the first to call for DuPont's corporate hide. But to date, despite all the accusations and rumblings, and after fifty years of C8 use in making Teflon at its Washington Works facility, that is* not *the case.*
>
> *Until cooler heads prevail and the facts come out, it's way too early to jump to conclusions—whether you're a judge or a regular citizen.*

The message being fed to the public was clear: A bunch of greedy plaintiffs' lawyers and an incompetent judge were set on destroying the local economy. If unchecked, they would ruin not only the town but everyone else's way of life.

Internally, DuPont was gearing up to launch a $20 million consumer-brand advertising campaign showcasing new Teflon-treated fabrics for clothing, household textiles, and carpets.

DuPont even called in the chief—CEO Charles "Chad" Holliday—to publicly defend PFOA and Teflon. He opened the annual shareholder meeting by trying to assuage shareholders' concerns: "We have not seen any negative effects on human health or the environment at the levels of exposure at which we operate." There was that phrase again. But coming from the boss's mouth, it carried added significance. Holliday was now personally defending Teflon and PFOA—on camera, in public, on the record, and in written SEC filings. His statements sounded confident and unequivocal. Backed by the reputation of "the science company," they seemed pretty convincing.

To someone else, maybe. I was convinced only that DuPont was becoming increasingly desperate. Getting personally involved was a risky move for a CEO. By doing so, he had abraded the nearly bulletproof case law that often protects high-level executives from being dragged into litigation. I realized that his public statements had opened the door for me to request his emails and any PFOA-related documents on his computer or in his files.

They also gave me grounds to ask for his deposition. It was a long shot, sure. Deposing a CEO is extremely rare—almost unheard of. But DuPont had opened the door by propping him up in front of shareholders, the press, and the public to speak about the safety of PFOA. After his speech at the meeting, I presented the court with the argument that his personal involvement now overrode the protective case law. If he was fit to speak about these issues in public, he could do so on the record—in a formal deposition.

The court agreed.

. . .

Before I got the chance to take Chad Holliday's deposition, DuPont made another astonishing move: they filed a motion to disqualify Judge Hill, arguing that he was himself a member of the class in our class action and therefore conflicted. As a resident of Parkersburg, they maintained, he was purportedly drinking "PFOA-contaminated water" at home. So DuPont's legal team was trying to get rid of the judge who had forced them to turn over all those documents to me—the ones that were finally fueling EPA's crackdown, the EWG's reports, and the bad press.

"The burden is on the judge, once he's determined that he has an interest in the matter, to voluntarily disqualify himself," DuPont's lawyer, Larry Janssen, said.

Even as I stood in the hearing before Judge Hill, who was just as flabbergasted as I was, I couldn't believe that DuPont had had the nerve to go after him. I'd recently requested access to Washington Works' original employee files, where I hoped to find the raw data and worker studies that had allegedly shown "no adverse effects." DuPont had refused. Judge Hill was the one who would be deciding: the same judge who had recently entered sanctions against the company, had ordered them to make blood testing available for the entire class, and was letting me take the deposition of the company's CEO. I thought the company must really, *really* want to keep me out of their workers' medical records.

There was also still the outstanding issue of the documents DuPont claimed were privileged but we claimed they had waived. We were locked in a stalemate that could be broken only by Judge Hill.

But Judge Hill couldn't rule on any of these issues—if he were disqualified.

How ironic that DuPont had fought to have our case moved to Parkersburg. So much for home-court advantage. Judge Hill's rulings had not gone the way DuPont wanted. So now they were trying to get rid of him.

It was a risky move.

"I was absolutely shocked and surprised by your motion," Judge Hill told DuPont's lawyers. "Because I had no idea that the people in Parkersburg were affected. There was never any talk about anyone except for those in Lubeck . . . and over in Ohio, the Little Hocking area. But not Parkersburg."

Judge Hill drank Parkersburg water. Parkersburg samples were beneath the laboratory quantification limit for PFOA in drinking water (0.05 ppb, according to a DuPont scientist). But DuPont's lawyers—who would later argue that anyone with water contamination below the quantification limit should *not* qualify as a class member—now found it convenient to argue that Parkersburg water customers *were* class members. Including Judge Hill.

"But everybody in the country apparently is affected," the judge said. "So, what in the world is going to be the limit? Are you going to disqualify every judge that ever is appointed in this case?"

A judge "should not refuse to sit when qualified any more than he should insist on sitting when disqualified," he said. "In other words . . . I can't get out of the case because it seems to be a burden—which it really is."

That was another thing we had in common.

Judge Hill denied the motion to disqualify.

He then proceeded to address the other pending motions. It was a clean sweep—against DuPont. He ordered the company to give me the plant's employee medical data, which DuPont had been fiercely resisting. He rejected DuPont's claims of privilege and held that they had to give us not only the specific documents in dispute but all related documents — including the discussion about what to tell the community about PFOA in their drinking water.

"You may ask the honorable members of the Supreme Court of Appeals if they agree with me," he said.

DuPont did just that. Again.

In the summer of 2003, the Supreme Court of Appeals of West Virginia ultimately decided that it would hear DuPont's appeals on three of Judge Hill's rulings: his blood-testing injunction, his decision not to disqualify himself, and the waiver of privilege.

These appeals stayed our case. Stopped us cold. All additional discovery and formal proceedings with the court beyond what it had already ordered in regard to the production of DuPont's internal medical records was frozen: no more motions, no depositions, no other forced discovery until the stay was lifted. Only a decision by the state's highest court could do that.

22

EPIDEMIOLOGY

July 1, 2003
Parkersburg, West Virginia

Additional formal proceedings in our case might be on hold, but I wasn't.

Freed from filing motions and taking depositions, I suddenly had a whole new resource: time. Now I had the time to attend EPA meetings in DC, held to collect public input into EPA's "priority review" of PFOA for potential regulation. I also had time to hunt down the court-ordered medical records, and acquire another foreign language: epidemiology. According to DuPont, all the toxicology data and test results showing serious harm to lab animals exposed to PFOA (including cancer in lab rats) were meaningless unless we could prove that any of the same effects occurred in humans. To do that, they claimed, we would need the results of actual studies of PFOA exposures and disease in people: epidemiologic data. With my foray into epidemiology, I was embarking on the most challenging and unique aspect of our legal strategy, something that had rarely if ever been attempted in similar cases. To prevail, I would not only need to use the scientific secrets I'd dug out from DuPont's files, I would have to go well beyond where DuPont had been willing to go, actually showing the impacts of PFOA on our class members. And I would have to do it with the already strained resources of Taft and my co-counsel's law firms, achieving something that DuPont, with their billions of dollars and high-paid staff of specialists, the state of West Virginia, and EPA, with all the power of the federal government, had failed to do.

I couldn't begin to do it on my own, of course. To help me get started, I found the perfect expert: Dr. James Dahlgren. Dahlgren was a medical doctor who had spent years studying the human health effects of industrial pollutants. Most famously, his research on the toxicity of hexavalent chromium, a corrosion-fighting chemical used by Pacific Gas and Electric that, like PFOA, had then been dumped into unlined ponds that had leached into drinking water, had helped decide the California pollution case portrayed in the film *Erin Brockovich*.

One of the first things Dahlgren explained to me when I contacted him was the critical gap that existed with respect to the human data DuPont had collected. DuPont had focused exclusively on their workers—who were primarily adult healthy males in good enough health to be working at the plant—and had not done anything to collect data on the exposed community surrounding the plant, which included children, the elderly, and the sick. We had asked Dahlgren to design a study to do just that. He had already begun collecting data from nearly six hundred residents out of the roughly seventy thousand class members and was in the process of sorting through and analyzing that information.

But we faced an even more urgent challenge. As I had been studying DuPont's summaries of their workers' health data, I had noticed a puzzling anomaly. According to the data summaries, Washington Works had had elevated rates of kidney cancer in 1989. In subsequent years, the problem had seemed to just go away. Dahlgren told me not to trust the summaries. If the data was incomplete, the summaries would be meaningless. Garbage in, garbage out. We needed to see the raw data. And to do that, we would have to access the individual medical files stored in the medical office at Washington Works. Judge Hill's ruling—before the case was stayed—that DuPont had to give me access to them was still in effect. Fortunately, it was not one of the rulings that DuPont had appealed to the state Supreme Court.

So on a warm July day Dahlgren flew in from Southern California to meet me at the plant. From an airplane, Washington Works could pass for a small downtown in a rural state: a cluster of buildings on gridded streets in a place where railroad tracks meet. It's a concrete island in a sea of green, covering a flat, clear swath of bottomland in a geologically crumpled state where the hills are furred over with trees. It's a blip on the 981-mile Ohio River, a little shy of the halfway point between Pittsburgh and Cincinnati. From the ground, it's a maze of concrete and metal, an industrial labyrinth that stretches a full mile along a river that sometimes sparkles and sometimes looks like Yoo-hoo. A matrix of pipes branches out at sharp angles throughout the plant, like rail lines on a subway map. They weave, veinlike, among giant cylindrical tanks and boxy, windowless buildings. Tendrils of vapor curl into the sky from smokestacks as slender as cigarettes. Though I had imagined it many times over the years, this was the first time I had actually walked in the shadows of the pipes and stacks. As I passed by plant employees, it struck me that all the decisions that shaped their lives, and possibly endangered them, were being made four hundred miles away in DuPont's gleaming high-rise offices in Wilmington, Delaware.

Dr. Dahlgren and I were taken to a room filled with banks of file cabinets. We had already agreed to maintain the confidentiality of the individual employees' health information, so our escorts left us alone to go through them. Working in tandem, we pulled out drawer after metal drawer and walked our fingers through the files, pulling out the folders containing the records of individual employees. We thumbed through everything, from on-site injury reports to correspondence among plant doctors, private physicians, and hospitals. We were looking for death certificates, records of illnesses, and cancer incidence reports. From my high-speed education in epidemiology, I understood that the medical files of these workers would reveal whether there was an elevated cancer rate at Washington Works as compared to the general population. If it was elevated, the files could be cross-checked with the workers' PFOA blood levels. It would be a critical step toward confirming that PFOA really did present a substantial risk to human health and that DuPont had been misrepresenting their workers' health data.

Every time we saw any reference to a cancer death or diagnosis, we noted it. There were a distressing number of them. As I'd suspected, when we hand counted the incidence of cancer cases among plant workers, they didn't match the numbers in the summary reports that DuPont had claimed indicated no unusual spike in the incidence of cancer.

As we worked, I made sure not to drink the water.

23

"NO KNOWN HUMAN HEALTH EFFECTS"

November 2003
Nationwide

Now five months into the Supreme Court stay with no end in sight, with Thanksgiving just a few weeks away, ABC News turned the spotlight on Teflon. As Americans had cooking and food on their minds, *20/20* host Barbara Walters gave them reasons to think—and worry—about the dangers of their cookware.

"It coats the pot you cook with so the food doesn't stick. It protects the carpet your baby crawls on. You may also have it in your winter jacket, skin lotion, even your makeup," she said. "We're talking about Teflon."

This was prime-time TV with an audience of 9.6 million Americans getting ready for the most food-centric holiday of the year. For DuPont, it seemed as though this PR disaster couldn't possibly get any worse.

And then they played the wedding march. The camera cut from Barbara Walters to a flower girl in a white dress. A bride and groom emerged from a small church, holding hands and beaming, ducking under a shower of rice tossed by guests.

Off camera, a correspondent intoned, "For the parents of the groom, this happy day is one they feared might never come, given the way he started life."

The next image was a yellowed photograph of a newborn baby asleep in someone's arms. The baby had one nostril and a red, misshapen eye.

More than two decades had passed since that photo had been taken. That baby was now the groom in the tux, a twenty-two-year-old man named Bucky Bailey.

Thirty surgeries had left Bucky with pink scars crisscrossing his face. Those surgeries had given him a second nostril and a better eye. But they had left a mark of their own. A two-inch scar rose between his eyebrows and diverged into a Y. That was where his forehead had been stretched to harvest skin to build a nose.

A succession of family photos told the story of his life. There he is, swaddled in a baby blanket. His right eye is an angry red mess, but his little pink lips are smiling. There he is, grinning and pushing a yellow exersaucer across a linoleum floor. A white patch covers his eye. Both arms are held straight by sleeves designed to prevent him from touching his eye. Then he's a little boy asleep in a hospital bed, his face swathed in gauze. Now he's a big kid in a baseball jersey, with a new nose and a bashful smile.

"I have never, ever felt normal," Bucky said, looking into the camera with one penetrating eye. "You can't feel normal when you walk outside and every single person looks at you."

For DuPont, Bucky was likely seen as a nuclear weapon. The best doctors in the world might never be able to prove that PFOA had caused his defects. But in the court of public opinion, that didn't matter. Watching, I could almost see inside the worried heads in DuPont headquarters: *Imagine if he ever got in front of a jury.*

I had made sure the *20/20* producers had whatever nonconfidential public documents and photos they needed. They had also asked DuPont for a spokesperson to put on camera. DuPont had chosen an executive with a science background to speak on behalf of the company. Dr. Uma Chowdhry, vice president of research, had a PhD in materials science from MIT.

"We are confident when we say that the facts, the scientific facts, demonstrate that the material is perfectly safe to use," she said in a robotic face-to-face interview with *20/20* investigative reporter Brian Ross.

But Ross did not just focus the public's concern about breathing fumes while frying eggs in their Teflon pans. He specifically asked about the PFOA used to make Teflon. That chemical, he made clear, was now showing up in everyone's blood. He asked Chowdhry about that.

"Everyone has it?" Ross said.

"Everyone has it," Chowdhry replied.

"It's in my blood? Your blood?"

"Possibly. We do not believe there are any adverse health effects."

"Is it a good thing to have it in your blood?"

"There are lots of chemicals that are present in our blood."

Bacon sizzled in a Teflon pan. Representatives from the EWG were using a special thermometer to test one of DuPont's claims: that pans don't get hot enough on a stove to trigger the release of toxic Teflon fumes.

DuPont had long acknowledged that fumes from overheated Teflon pans could kill birds and cause temporary flulike symptoms. But the company claimed that such issues would occur only if pans were heated well above 500 degrees Fahrenheit—a temperature it said pans don't exceed in normal home-cooking conditions.

During the segment, the EWG cooked bacon in a Teflon pan to address that claim. After a few minutes: "At 554 degrees Fahrenheit, studies show that very fine particles start coming off the pan. These are tiny little particles that can embed deeply into the lung," said Jane Houlihan, a spokeswoman for the EWG. "At 680, six toxic gases begin to come off of heated Teflon."

Brian Ross asked DuPont's Uma Chowdhry about that.

"We get some fumes, yes," she said. "You get flulike symptoms, which is reversible, and if you follow the instructions on the pan . . ."

"You feel like you have the flu?" Ross asked.

"You feel like you have the flu—*temporarily,*" Chowdhry said.

"How long does that last?"

"Temporary. Couple days."

"We cooked some bacon," Ross said. "Above five hundred degrees, the bacon still wasn't done."

You had to love Chowdry's response.

"I've never cooked bacon," she said. "I can't comment."

She did, however, comment on birth defects.

The producers of *20/20* included interviews with not only Sue Bailey but Karen Robinson. With brown eyes that worried behind wire-rimmed glasses, Karen spoke of her son, Chip, born with an eye defect, and her daughter, who had originally appeared to have been born healthy.

"Two years ago, we discovered that she has a birth defect that affects her kidneys," Karen said. "One kidney did not grow. One kidney grew to three times its normal size."

Like Sue Bailey, Karen Robinson was angry with DuPont. "DuPont should be held accountable for their actions in keeping all this secret from the public."

To that, DuPont's Uma Chowdhry responded, "In the general population, incidences of birth defects are not uncommon."

"Two out of eight workers is not uncommon?" Brian Ross asked. It was actually two out of seven births, but he had the right idea, and Chowdhry's answer would have been the same in any case.

"That was not a statistically valid sampling," she said.

"You study eight women who worked with PFOA. Two of them had children with birth defects. That would not be significant?"

"We had scientists pore over the data. In the realm of scientific fact, it is not considered a statistically significant sampling. All the other children were normal. And since then we have not seen a preponderance of birth defects."

"Have you done a study to see?"

"No. We have not."

The show closed with another wedding scene of Bucky and his new bride walking between packed pews. He is dressed in a black tuxedo with a white rose tucked in his lapel. He is smiling.

On camera, in a later interview, Bucky is not smiling. His eyes are shiny with tears. His voice cracks when he talks about his future. "I have to think about whether I want to have children or not. I cannot put them through what I went through."

. . .

The *20/20* exposé was broadcast during peak holiday shopping season, when many consumers would likely be considering a purchase of nonstick cookware. If the show diminished consumer trust in the brand, it might be hard for DuPont to regain that trust.

But in December 2003, DuPont received a consolation prize: the Supreme Court of Appeals of West Virginia came back with a ruling on the first of DuPont's three appeals.

Judge Hill's injunction was thrown out on a technicality. DuPont would not be forced to make blood testing available for our seventy-thousand-member class. It was a major win for DuPont, not only because of the money it would save on the testing. Those blood tests could have generated the critical "significant exposure" data DuPont said was needed to prove our medical-monitoring claims, since we refused to play along with the CAT Team's absurd 150 ppb guideline.

Meanwhile, Dahlgren had been busy analyzing his new data from our class members. His epidemiological survey indicated that the incidence of all cancer types (excluding skin cancer) was 8.65 percent in those exposed to Washington Works' PFOA, compared to less than half that—3.43 percent—in the general population. Dahlgren found what he judged to be statistically significant increases in uterine/cervical cancer, multiple myeloma, lung cancer, non-Hodgkin's lymphoma, and cancers of the bladder, prostate, and colon/rectum. He released a summary of his findings at a scientific conference, at which point they became public, and I promptly forwarded them to the public EPA docket, noting their relevance to the ongoing public-health threat from PFOA. Not surprisingly, the conclusions made a splash. A May 2004 headline from Ken Ward in the *Charleston Gazette*: "Study Links Teflon Chemical to Higher Cancer Rates."

Now not only was PFOA associated with "health effects," it was being mentioned in the same breath as something far scarier: cancer.

DuPont swiftly attacked Dr. Dahlgren and dismissed his study as "junk science." The headline on the front page of the *Parkersburg News and Sentinel*: "New Study Finds Cancer Rate Higher in C8-Exposed Areas: DuPont Refutes Report, Questions Validity."

The story quoted Haskell Labs' Bobby Rickard steadfastly singing the familiar refrain: "PFOA is not a human carcinogen, and there are no known health effects associated with PFOA." He dismissed Dahlgren's claims as "inaccurate" and "inconsistent with published scientific studies." He scoffed at Dahlgren's conclusions and called the methodology "unscientific." "In fact, the more we study PFOA," he said, "the more confident we are in our conclusions that PFOA is safe."

With the state's highest court ruling in their favor on the blood-testing issue, DuPont once again went on the offensive. Around the time Dr. Dahlgren's results were making headlines, DuPont issued a press release announcing their plan to do a new worker study at Washington Works. This one would be a million-dollar study aimed at "determining whether a substance used in the manufacture of Teflon causes any adverse effect."

The science war was escalating. DuPont would fight data with data.

24

CORPORATE KNOWLEDGE

As Teflon kept making news, lawyers on both sides of our case were busy preparing for trial—delayed multiple times but now set for October 4, 2004, seven months off. No longer completely stayed, DuPont was deposing our scientific experts. I was countering by aiming for the head—deposing DuPont's CEO.

I flew to Wilmington, Delaware, and checked into the Hotel Du Pont the night before. As I was about to leave for the airport, my parents called. My mom was on one extension, and my dad was on the other. Both were as nervous as cats in a canoe and peppering me with questions: "You're staying in the *Du Pont* Hotel? Does it have to be *there*? Who else knows where you'll be? Will anyone be with you? Are you sure it's safe?"

I told my parents that the biggest threat to me would be the price of minibar items. But that night I slept poorly. As I lay awake shifting uncomfortably in the hotel bed, the implications began to multiply in my mind. The DuPonters knew that everything I'd discovered about their long and culpable history with PFOA left them severely exposed. They'd tried to derail me from the beginning, first playing me for a fool, then stonewalling me on discovery, and when none of that had worked, they'd even tried to get rid of the judge and stop the whole case. What made me think they would stop there? Was the phone in my hotel room bugged? Was I being watched? Could anyone mean me harm?

I became painfully aware that most of the complicated case against DuPont was known only to me. I had laid out the facts in my letters to EPA and continually briefed our legal team, but making sense of it all wasn't easy. There were so many moving parts, so many gears of law, science, and history that had to mesh in just the right order to make the case. If somehow I fell out of the picture, the whole issue might just disappear.

After calling Sarah to wish her good night, I settled into my New College hoodie with its permanent buttered-popcorn stains on the front and my flannel sleep pants covered with Santa faces and the words "HoHoHo!" on them (affectionately called my hoho pants), and turned on the Cartoon Network for some relaxation. But it never came. I'm always hot, so I finally cranked the room's air conditioner way down and set the bathroom exhaust fan on high in an attempt to create enough white noise to make the room sleep-worthy. It didn't work.

The next morning, I shook off the lack of sleep and comforted myself with routine. My boxes crammed with cross-referenced documents had arrived at the hotel the previous day, sent via FedEx by Kathleen Welch, my paralegal. I had picked them up, hefted them onto the folding hand cart I'd stowed in the overhead compartment on the plane, and rolled them onto the elevator and down the long hallway to my room. I'd worked late rummaging through the boxes, rereading each document, annotating it with a highlighter and a pen, then placing it in the order in which I would refer to it in the deposition. After a cup of hotel coffee, a bit stronger than I liked it, I put the boxes back onto the handcart and trudged back down the hall to the elevator. The deposition, set to start at 9:00 a.m., was on another floor in one of the hotel conference rooms. I arrived a few minutes early to find the court reporter setting up her stenography equipment and the videographer laying a cable down the middle of a long dark wood table and setting up a blue backdrop like the kind they use at an Olan Mills portrait studio. The walls were dark green, a hue you might expect in an old English gentlemen's club. The overall feel was a bit claustrophobic.

Holliday would sit at the end of the table in front of the backdrop.

The court reporter would sit in the chair immediately to his left, and I would sit beside the court reporter. DuPont's attorneys would face us across the table. During the deposition, which could last up to seven hours, not including a lunch break, only one of DuPont's lawyers would be allowed to say anything. That would be Janssen, their lead attorney, who had yet to look me in the eye.

I was finishing organizing my documents when the DuPont lawyers rolled in, shaking hands and doling out business cards to the court reporter and videographer. Janssen, as always, looked right past me. Holliday entered last, wearing an expensive suit and a seemingly genuine smile. He had abundant sandy blond hair brushed back off his forehead and a handsome face gone slightly fleshy, especially around the neck. He lit up the dark room with confident charm. His bonhomie wouldn't have seemed out of place at a cocktail reception. He sat down in the hot seat, and the videographer set up his microphone. The court reporter asked if we were all ready to begin. I was more than ready. I'd been waiting a long time for this.

Chad Holliday had been at the helm of DuPont since 1998, the year Earl Tennant had first called me about his dying cattle. That was the same year the tobacco industry had settled lawsuits filed against it by fifty states for $246 billion. Holliday would have to steer the company through a new era of corporate responsibility—and liability.

He took on those challenges forthrightly and aggressively. He was the sitting chairman of the World Business Council for Sustainable Development. He had authored a book called *Walking the Talk: The Business Case for Sustainable Development*, in which he had made a case for sustainable development and corporate responsibility.

He seemed like a new breed of CEO, enshrining the quest for social responsibility as a prime directive. It all sounded so sincere, but what about PFOA?

On that subject, it was the same old line: "We have not seen any negative effects on human health or the environment at the levels of exposure at which we operate," he told investors.

In fact, it was under Holliday's leadership that DuPont had decided to start making PFOA. Within months of 3M's announcement back in May 2000 that it would discontinue manufacturing PFOS and related products, DuPont was hashing out plans to build their own PFOA production line in Fayetteville, North Carolina. That was a decision I wanted to home in on: just as 3M decides a chemical is too risky to produce, DuPont jumps in headfirst? I asked Holliday why the company had decided to manufacture PFOA.

"After weighing all the factors we weighed, hearing all the scientific evidence and all the opinions, we made the decision it was the right thing to do to produce that product."

The right thing to do? That still didn't jibe.

I pulled out DuPont's annual report and read him the statement he had made to investors: "Based on over fifty years of industry experience and extensive scientific study, DuPont believes there is no evidence that PFOA causes any adverse health effects or harms the environment."

I was so tired of hearing those hollow words, that universal reply to any question about PFOA. I could only imagine the team of lawyers, executives, and PR consultants it had taken to craft it. In fact, I'd found a document that revealed its desperate importance. It was an email from DuPont public relations director Diane Shomper to the lawyers:

> The no observed health effects and no evidence of human health impact are the basis of our whole public position and as far as I know are true. What do we do without that?

Holliday, too, said he believed these statements to be true. *As far as he knew.*

That's when my nagging sense that there was something revealing in the careful wording of DuPont's mantra finally resolved in my mind. Suddenly I understood why the wording had rung that faint and for so long uninterpretable alarm in my brain. *We have seen no evidence. As far as I know.* Those at first innocuous phrases shoehorned into the bold exclamations of NO EVIDENCE rendered the conclusion meaningless. How can you "see" evidence if you don't look for it—or if you avert your gaze? Just because you don't know something doesn't mean it doesn't exist.

I read Holliday another statement he had made in the annual report: "DuPont does not believe that consumption of drinking water with low levels of PFOA has caused or will cause deleterious health effects."

Did he stand behind that statement? Even now? Under oath?

"I have no reason to question that this is not correct," he said.

I didn't think he was lying. Holliday came across like a man who says what he means and means what he says—a man who truly believed in "walking the talk," as he had written in his book. But I was starting to get the impression that he "had no reason" to believe it because nobody had given him a reason. He didn't have all the facts and hadn't bothered to seek them out.

I wondered—as I'd wondered about Paul Bossert, the plant manager who had earnestly defended PFOA at the town hall meeting in Little Hocking—if Holliday had been deliberately kept in the dark. If so, that would explain how he could stand up and speak with conviction about the "safety" of PFOA. He just didn't seem like the kind of guy who would stand there and lie through a smile.

The boxes I wheeled in on the handcart contained more than a hundred documents—each copied in triplicate. One had my notes on it. The second set was for DuPont's attorney. The third was to hand to Holliday in the course of his deposition. Once they were recorded as exhibits, he could not say he'd never seen them. Nearly all of the exhibits, of course, had been pulled directly from DuPont's files.

As I showed Holliday page after page of evidence contrary to his statements, evidence produced by his own company, his confidence seemed to slacken. With every question I asked, it became that much clearer—to both of us—how much he did not know.

What did he know about the rat birth-defect study on PFOA?

"I'm not aware of that study."

What about the 1981 data showing PFOA in the blood of female plant workers—at levels above the 400 ppb that DuPont's scientists said would mandate their removal from the workplace?

"I'm not familiar with that data."

And the fact that those women had had babies with PFOA in their blood? Did he know that PFOA had been found in umbilical cord blood? That it crossed the placenta?

"No."

The birth defects? Among babies born to female Teflon workers?

"I'm not aware of those claims."

Did he even know about DuPont's own 1980s data showing PFOA in the public drinking water? The data DuPont had not disclosed to the community?

"No."

Well, now he did.

In fact, he now knew a whole lot more about PFOA than when he had walked into the room that morning. He had acquired more new and useful information from his deposition than I had. That was the whole point of the deposition. I didn't expect to learn anything I didn't already know. I wanted to make sure that Holliday could no longer stand up and say, "I have seen no evidence of health effects." Because now he had seen it.

After the cameras were off and the court reporter stopped typing, we packed up to leave. On the way out, Holliday turned to me, extended his hand, and looked me square in the eye. I saw no animosity.

"Thank you," he said, shaking my hand.

I believe he actually meant it.

• • •

The deposition ended at five. I went back to my room, got my bag, and went straight to my rental car for the drive to the airport. I had been disarmed by the look in Holliday's eyes when he had shaken my hand. But had anything really changed? In the five years I'd been litigating against DuPont, it was the first time I hadn't felt that I was dealing with a win-at-any-cost mentality so powerful that it could lead normally moral people into making decisions that could put others at risk. Had my parents' bout of paranoia been all that unreasonable? I could hear my father saying "Be careful, son. Maybe you should make sure you aren't the only one who knows all these things," and immediately my mood shifted back to the unease of the previous night. Maybe I'd read too many legal thrillers, or maybe I was just exhausted from the mental strain of conducting the deposition. But as I slid behind the wheel, I paused and took a shaky breath before I turned the key.

25

THE PERFECT STORM

In June 2004, new litigation against DuPont escalated the crisis. The new claim was not filed by an exposed individual, a business entity, or even a class. This time the plaintiff suing DuPont was EPA.

Despite pledges from DuPont a year earlier to work with EPA in "good faith to help the agency with its 'priority review' of PFOA," the agency's efforts to get the company to supply complete data on nationwide exposure sources of PFOA had gone nowhere. It seemed that EPA was finally calling out DuPont's "good-faith" effort as a sham, a ploy designed to buy time and fend off formal regulatory action while "negotiations" continued indefinitely. I had been traveling back and forth to DC to attend all those public meetings sponsored by EPA, representing our class members, and had witnessed firsthand the hours and hours of endless talk that went nowhere, DuPont simply spinning the data and buying time.

EPA's suit went after DuPont for failure to report the research I had sent EPA from DuPont's own files showing that PFOA had been transmitted from a pregnant DuPont worker to her fetus and that it had been found in public drinking water near DuPont plants. It was a dramatic statement and proved to me that, at least for the moment, the government was climbing out of DuPont's pocket. This was one of the few instances in history when EPA had sued a corporation for failing to report chemical toxicity and health-risk information. EPA claimed that DuPont had broken federal laws. And what was the evidence it brought? As I flipped through the document, I saw five years of my life flashing by: all the toxicity data and water-testing studies I had sent EPA from the 1980s that revealed what DuPont had known—and chosen not to report—to EPA. It was using the blueprint I had painstakingly provided to it starting with the twelve-pound letter in 2001, bolstered by three years' worth of scientific data that I'd been steadily passing along.

It had taken a while, but this complaint was proof that, finally, the EPA folks who had taken my thick letters seriously had prevailed. My letters were cited in the complaint as a source of their information. All those hours spent reading, the late nights at the office, the time I'd missed with my wife and children, the constant stress of worrying about DuPont's next attempt to derail me and my lack of financial productivity for my firm—it all added up to this.

The impact of my efforts continued to snowball. EPA, under increasingly intense pressure from the media, the EWG, and me, had finally taken some real enforcement action to show the world it was doing something. I didn't have access to EPA's internal deliberation notes, so I couldn't know what had triggered its decision to move against DuPont now. Had it simply come around, or was it protecting its own backside? With so much public scrutiny on the issue of PFOA taking place, surely it wouldn't take long before people started asking why EPA had let all this happen in the first place. Was the agency simply getting ahead of the issue before it bit it, or had its eyes truly been opened to the dangers of PFOA? Like everyone else, I could only guess. Around the same time, the Little Hocking Water Association, to reduce its own liability, sent a notice to its customers referring to the presence of PFOA in their water and warning them that they now drank the water "at their own risk." The effect of all this in the public mind was almost as toxic as the water. As one Parkersburg TV story put it: "Doom for DuPont?"

EPA's complaint alleged a total of eight violations beginning as early as 1981. There was buzz about its potentially generating "the largest fine ever." By some reports, EPA officials could fine DuPont more than

$300 million, ten times more than the maximum fine the agency had ever assessed to that date.

I knew from my years working for companies being sued by EPA that the agency often ended up resolving whatever kind of enforcement case it brought for far less than the possible maximum fine. But the immediate cost of EPA's complaint was another hailstorm of bad press—this time around the world. And in at least one instance, that press coverage created a public-relations catastrophe. Stories in Chinese media characterized consumer response as "mass panic" and reported that sales of China-made brands of Teflon cookware had fallen by more than 60 percent in one week. The response was so alarming that Chad Holliday flew to Asia to attempt damage control.

EPA's complaint alleged that DuPont had actively tried to conceal information about the dangers of PFOA. Suddenly, the idea of a cover-up no longer sounded like conspiracy theory. Holding DuPont responsible for their actions had always been one of our goals in the legal process— the one Earl had felt most passionately about. Now the federal government was finally acknowledging that DuPont had acted wrongfully. That was profoundly satisfying. It also reflected well on our class action suit and elevated our standing in the media and legal community. As gratifying as that was, it still didn't accomplish our primary goals of protecting our class members through medical monitoring and a court-ordered supply of clean, safe water. It was a win, but it was nowhere near the victory we still needed to fight for.

Pressure on DuPont was mounting on all fronts. The PFOA story was going global. The cover-up had been exposed. The regulators were knocking at the door. Teflon sales—at least in China—were tanking. And the public's trust was now in question.

If things didn't seem bleak enough already for DuPont, the Supreme Court of Appeals of West Virginia finally came back with a ruling on the last two of the company's appeals. DuPont's legal team must have seen both as crushing defeats. Their attempt to force Judge Hill to disqualify himself from the case had been rejected. And while the court was at it, it had let stand his ruling on DuPont's waiver of privilege on certain internal lawyer communications we had gotten through discovery. Du-Pont had claimed that their giving us messages from and between their lawyers had been a simple mistake, and in a nonlegal sense it certainly had been a mistake—a colossal one. But Judge Hill had ruled that, in a legal sense, the documents had been sent to us in a deliberate way that contradicted the "mistake" defense. In the end, the Supreme Court let stand Judge Hill's ruling that DuPont had waived privilege not only over specific documents but over the basic subject matter of those documents. That freed us not only to use those documents but to seek all the other documents concerning the same issues that had been claimed privileged throughout the litigation, dating back to the original Tennant case. This was huge.

One of the documents DuPont had produced and been fighting to get back was a November 8, 2000, email authored by in-house litigation attorney John Bowman to his superiors in legal, including the top lawyer at DuPont at the time, General Counsel Tom Sager. Bowman had shared his concern over DuPont's continuing to stonewall on the issue of contaminating the water downstream from Parkersburg. He had sent the email nine days after the October 30, 2000, letter mailed to Lubeck water customers that had finally disclosed the presence of PFOA in their water, ten years after the fact. I also noticed that it had been sent not long before he'd had that odd discussion with Larry and me after our deposition of DuPont's water-testing expert.

I was amazed to see that Bowman was advocating that DuPont fulfill one of the key goals of our suit: "I think we need to make more of an effort to get the business to look into what we can do to get the Lubeck community a clean source of water, or filter the C-8 out of the water," he wrote. He also expressed extreme pessimism about how the issue would fare in court.

"My gut tells me the biopersistence issue will kill us . . ." he continued. "We are going to spend millions to defend these lawsuits and have the additional threat of punitive damages hanging over our head."

And then there was the amazing revelation: "Bernie [Reilly] and I have been unsuccessful in even engaging the clients [DuPont] in any meaningful discussion of the subject. Our story is not a good one, we continued to increase our emissions into the river in spite of internal commitments to reduce or eliminate the release of this chemical into the community."

This email revealed that people inside DuPont were pushing the company to do the right thing, the most primary, basic, and doable thing: get the community clean water to drink. If nothing else, that one thing should have been nonnegotiable. The lawyers had made that clear.

And the company had apparently ignored them.

This was huge—not just in a moral sense but in a legal one.

This would help nail the argument for punitive damages. If the memo wasn't evidence of conscious disregard, I don't know what would have been. DuPont's own in-house attorney was acknowledging that the facts in this case could leave the window wide open for punitive damages that could far exceed even those of actual damages.

I was so relieved that, for the first time in months, I didn't go home dreading a return to the office later that night or the next morning. I played extra rounds of "bag, train, monster"—the boys' favorite game, which involved my swinging them around in a blanket ("bag"), pulling them around the house on the blanket ("train"), and hiding under the blanket ("monster"). After the boys were in bed and Sarah had poured me a celebratory glass of wine, I began telling her the significance of the documents we could now use.

I suspected that that was why DuPont had battled so long to keep these documents under wraps. Even I could work up a little sympathy imagining poor John Bowman picking up his *New York Times* on Sunday, August 4, 2004, to find a screen grab of his instantly infamous memo plastered on the front of the business section. After all, it seemed he had been desperately trying to get his client to do the right thing. Once the state Supreme Court ruled that Judge Hill's privilege-waiver ruling would stand, the copies of the documents that had been submitted to the court under seal pending appeal were unsealed and became public record, available to anyone accessing the court records—in other words, fair game for the *New York Times* reporter.

The *Times* story wasn't the only nasty surprise waiting for Bowman. Not only did the state Supreme Court ruling in our favor entitle us to all those juicy documents, it meant we could question the people who had produced them. Like CEOs, attorneys are typically protected from being deposed. But just as Chad Holliday's personal statements had undercut that protection, the court's refusal to make us give back the emails we had received in discovery from Bowman and Reilly made them fair game for interrogation. I sent out formal requests for both of their depositions and the rest of their files on the subjects over which the privilege had been waived.

The call from DuPont's legal department came through to my office early the next day, September 1. Would I agree to "postpone" the depositions in exchange for DuPont agreeing to mediation talks?

I felt no sense of celebration. All earlier mediation had basically been a waste of time. This time, though, I knew the company wasn't just stalling. The idea that their own lawyers might have to appear in court as witnesses to defend their documents seemed to have lit a fire under them. Mediation might be their only way to mitigate the damage. It also might be our best opportunity to do what we'd been promising for far too long: to bring some relief to a whole bunch of people drinking tainted water.

26

THE BIG IDEA

September 4, 2004
Boston, Massachusetts

W e gathered in Boston for a knock-down, drag-out mediation.
It was all hands on deck: Larry Winter, Harry Deitzler, and one of Harry's and Ed's other partners, Jim Peterson, flew up from Charleston. Joe and Darlene Kiger, two of our lead plaintiffs, represented the class. One of my firm's senior partners, Gerald Rapien, came with me. By the time Ed Hill arrived from Washington, DC, serious negotiations were under way.

We were huddled in a conference room at the downtown Boston offices of Eric Green, a nationally known mediator specializing in highly complex cases. DuPont's pick, Green, had been named comediator with our choice, Jim Lamp, a sharp West Virginia lawyer who knew the state laws inside and out. The two men were the vehicles of our "shuttle diplomacy," relaying messages, both verbal and written, back and forth between us and a separate conference room where DuPont's legal team was strategizing.

The fact that we were physically separated for much of the session did two things: it permitted each side to communicate freely and confidentially among themselves, and it also prevented face-to-face conflict, which could easily have escalated. When we did finally gather in the same space, it was because we were nearing agreement and needed to hash out the details. But that was a long time in coming. Even within our own group we had different personalities and different negotiating strategies. Those differences made reaching consensus a contentious and often intense process.

Mediation talks can last as long as the parties believe agreement can still be reached, which can be minutes or days. In this case, we worked for almost eighteen hours spread over two days. It didn't help that the room was uncomfortably warm, a discomfort that increased with every hour. When mediation works—and it often does—it saves all parties the considerable time, work, and expense of a trial. Plus, any trial verdict is appealable, with the prospect of even more time, work, and expense.

All previous attempts at mediation had spun us in circles and gotten us exactly nowhere. This time, though, I was hopeful that we would reach a conclusion. How the case might play out before a jury was much clearer now than it had been. DuPont now knew that they risked having their own emails containing damning admissions, many from their own lawyers, projected onto a big screen in the courtroom. In deposing our witnesses and seeing our experts' opinions, the company had realized how much we knew and what the jury would be hearing. We had built a solid case. We had clear evidence for all the elements required for medical monitoring. DuPont's long history of water and blood testing proved high exposure. The record we'd uncovered of animal tests and worker studies demonstrated toxicity that could lead to disease. The anxious memos from DuPont's legal department presented a clear picture of management's wrongful conduct and even conscious disregard for the health and safety of those downstream. I was not about to agree to anything that felt like a capitulation. Our hand was strong. If we couldn't get a settlement that met our goals, I was perfectly content to continue down the path to trial on the class action claims for medical monitoring and clean water, however long that might take, and I'm sure DuPont knew that. In the court of public opinion, DuPont had already been injured greatly. Every day our case continued was another day that more damaging internal DuPont documents could end up in a public court file, in an EPA docket, and eventually in the press. Perhaps more urgently to their directors and investors, the company's sales and stock price had both taken hits.

Though I was well prepared for a trial, I knew we could get help for our clients much faster through a settlement negotiated through mediation. The most basic, fundamental right I'd been fighting for since the very beginning—clean water for our class to drink—was years overdue. That one simple, straightforward thing about this PFOA mess had never been addressed. For four years people had endured additional exposure because DuPont wouldn't listen to John Bowman saying "Get the Lubeck community a clean source of water or filter the C8 out of the water."

After clean water, what the members of our class action suit wanted most was answers. Was PFOA in their blood? If so, would it cause the same effects in them as seen in the lab animals and workers? What could they do to protect themselves?

Simple questions. Not so simple getting answers. DuPont had gone from not even admitting PFOA's existence to admitting its existence but claiming it was perfectly safe to saying that even if PFOA was really the cause of the health effects we pointed to in the lab animals and workers, the "trace" levels our class members downstream were exposed to in their drinking water were far too low to be cause for any concern. Dahlgren's initial epidemiology results certainly suggested that that reasoning was flawed, but no more comprehensive community study had ever been done. This was the weakest part of our case, not because of any science on DuPont's side but because of the lack of published, peer-reviewed science on community-level exposures. The extreme difficulty of remedying this gap, of generating the science that would confirm that the level of PFOA exposure suffered by the members of our class was, in fact, capable of causing some specific disease, was exactly what I believed DuPont was counting on and the reason they had fought us every step of the way, waiting for us to accept the ultimate futility of our case and go away. Their scientists knew how unlikely it was that links between PFOA and specific diseases in the community could be more definitively confirmed. The links almost certainly existed, but generating "statistically significant" values for diseases that strike only a small percentage of the population is devilishly difficult. As we had learned with Dahlgren's study, a sampling of hundreds of people would be viewed as nowhere near enough. And given DuPont's dismissal of results from their own worker studies, even a pool of thousands of people probably wouldn't be seen as large enough. Tens of thousands would likely be required. DuPont knew that no such studies had been undertaken and were probably quite confident that we would be unable to create one from scratch.

· · ·

At the end of the day, all settlements boil down to a number—a dollar figure. How much is the company willing to pay to make the problem go away? But a check would not fix our class members' biggest problem: PFOA would still be in their water and their blood. The full extent of the long-term consequences of that exposure throughout the community would still be unknown. The fear and uncertainty surrounding PFOA would not go away. No amount of money could answer all the lingering questions.

As the mediators shuttled back and forth bearing proposals, demands, and counteroffers, we focused on the problem that transcended everything else. Though money couldn't buy good health, it could pay for measures to improve and protect the health of seventy thousand class members—and the generations that would follow.

It could buy industrial-sized water filters for the water districts, so they would no longer have to warn their customers to "drink at your own risk." It could pay for similar filters in the homes of people who used their own private wells. It could pay for medical monitoring to detect and treat diseases early, when the odds of recovery are better. And it could fund scientific research to confirm what specific levels of long-term exposures to PFOA had actually done and could do to the people in the community.

The process was equal parts tedious and stressful. This wasn't speed chess. There were no clocks limiting deliberations between moves. Either we were just sitting as minutes dragged stubbornly by, waiting for DuPont to react, or we were trying to calculate how to respond, how much to demand, and how much to compromise. We were only too conscious of the years of effort and thousands of human lives at stake, and we felt torn between the desire to get a deal done now before more people in the class got sick and the determination not to yield. Because the substance of the discussions is strictly confidential, I can't go into specifics, but I can tell you that as the hours ticked by, the room felt increasingly stifling, and sweat beads formed where my collar cinched my neck. As the originator of and principal actor in the drama of the lawsuit, I knew that my opinion had added weight, but I had no formal authority to dictate. Ultimately, we all had to agree.

By the end of a long and harrowing day, we had reached a preliminary settlement. It included immediate, tangible relief (clean water to the class) and also a more subtle form of relief on the issue of giving people the answers they craved about what risks they faced after drinking tainted water for all those years. This was the heart of our settlement, something that as far as we could determine had never been done before: the creation of an independent "Science Panel" to confirm and document exactly what PFOA could do to those in the community who were drinking it. To ensure that neither side could later argue that the panel was biased, we would work with DuPont to jointly select three of the world's best epidemiologists, candidates who were not affiliated with one side or the other—whether through prior paid testimony, prior litigation, publications, or otherwise. Both sides had to agree that each candidate was properly qualified and unbiased. And either party could veto a candidate for any reason. The Science Panel would be charged with confirming probable links between PFOA and specific illnesses in the class. They would have to agree to devote themselves indefinitely to the work. The panel members' compensation would basically be whatever they requested. They would be given unlimited time and funds to conduct whatever research they deemed fit and to consult with whatever subspecialty experts they needed (subject to the same bias and veto rules).

DuPont would be required to pay whatever the study and panel cost, which, at the time of the settlement, was estimated at $5 million. The Science Panel would examine the risks of PFOA exposure among class members, who would by definition need to have had a minimum dose of at least 0.05 ppb PFOA—the minimum amount that could be accurately quantified in their water at that time—for at least one year. We agreed that the Science Panel could consider any data it deemed credible and sufficient—whether published, peer reviewed, or not—and that either side was free to submit whatever data it wanted to the panel, as long as the other side was copied on everything sent.

For another, more immediate form of relief, DuPont agreed to pay $70 million in cash "for class benefit," which could include direct payment to class members, though at least $20 million of it was to be used for "health and education projects" that would benefit the class. These projects weren't specified, but we knew that in a community facing increased but as yet unconfirmed health risks from years of drinking contaminated water, there would be a need to educate class members about how best to respond to those risks.

To clean the water, DuPont would pay for the design, procurement, and installation of state-of-the-art water treatment technology for all the drinking water sources contaminated with PFOA (estimated at the time to be a $10 million project). By the time of the settlement, the extensive water testing required by the 2001 consent agreement between DuPont and the state had found PFOA contamination in public water supplies in Little Hocking, Belpre, Pomeroy, and Tuppers Plains–Chester in Ohio and in Lubeck and Mason County in West Virginia, as well as in dozens of private wells in both states. Under our new settlement, DuPont would have to maintain the filtration systems at least through the end of the Science Panel's work. If the panel failed to confirm any links between PFOA and disease, DuPont could stop paying for the operation and maintenance of the systems. But if the panel confirmed a link with any disease, DuPont would be obligated to continue paying the filtration costs indefinitely.

Confirmation by the panel of any probable link between PFOA exposure and disease among class members would trigger the creation of a separate Medical Panel consisting of three doctors selected in the same way as the Science Panelists. These independent medical doctors would be charged with accepting whatever the Science Panel found and then determining what medical testing was appropriate to detect the linked disease(s) early, based on those findings. DuPont would pay whatever the Medical Panel's research cost, as well as for whatever medical monitoring the doctors ultimately recommended, up to a cap of $235 million. If no links to disease were confirmed by the Science Panel, there would be no Medical Panel and no medical monitoring for anyone.

While all that work was going on, all the class members' potential, individual injury claims based on PFOA in their water would be preserved and put on hold pending the outcome of the science. No one would lose a claim while the work was under way, but no one could actively pursue a claim in court against DuPont during that time. If the Science Panel confirmed that a particular disease was linked with the PFOA in the water, all class members who had that disease would then be able to bring individual lawsuits against DuPont for any injuries or damages associated with that disease, including for punitive damages. Claims arising from all the diseases for which the Panel did *not* find a probable link would be dismissed "with prejudice," which meant that no further claims against DuPont that those diseases had been caused by PFOA exposure from the Washington Works plant could be pursued by class members, even if future research proved the Science Panel wrong.

This arrangement was a gamble for both sides, but based on the evidence we had seen so far, we felt confident that, given sufficient data, the Science Panel would confirm links between class members' exposures to PFOA and disease. But confidence wasn't certainty. The risk that we were wrong was balanced by the fact that if we hadn't agreed to accept the Science Panel findings as final and uncontestable, DuPont could go on disputing the panel's findings forever and nobody would be likely to get compensation for anything—ever.

But how much evidence would the Science Panel need to find a probable link? We knew we needed to resolve that issue now, because if we had to argue about it later, we'd be back where we started. We eventually got DuPont to agree that the standard used in West Virginia's medical-monitoring law—the law that had provided the original basis for this whole case—would prevail. Under that law, "probable links" would be defined as those where, "based upon the weight of the available scientific evidence, it is more likely than not that there is a link between exposure to PFOA and a particular Human Disease among Class Members."

Perhaps the most critical aspect of our agreement was that the Science Panel would stand in the role of judge and jury on the critical issue of "general causation"—meaning whether the "community levels of exposure" to PFOA were actually capable of causing disease. In other words, if the Science Panel confirmed a "probable link," DuPont could not argue that any class members' exposure to PFOA was insufficient to have caused that disease. The Science Panel's probable-link decision would be the last word on the general causation issue for every class member. There would be no further arguing on that point—and no appeals for either side. We would settle this with science—science protected against bias and slant by the requirement that all Science Panel members had to be approved by both sides. Only impartial science could implicate—or exonerate— PFOA. Only science could answer our class members' questions. Either it would help community members understand and deal with the risks and effects of PFOA, or it would show that their exposures were not high enough to have caused them any harm.

With the CAT Team episode still giving me nightmares, I felt immense satisfaction that we had finally agreed to accept completely independent, unbiased, scientific answers—not hired-gun advocacy. I suspect that DuPont agreed to the elaborate plan at least in part because it seemed to present little threat to the company, as the current lack of community-level data favored them and they knew how unlikely we were to be able to come up with enough new data from enough of those community members to be able to confirm probable links with disease.

The final part of the settlement concerned our fees. We insisted that none of the attorneys' fees come out of any of the awards or benefits to class members. DuPont would have to pay any and all attorneys' fees and expenses separately and on top of whatever the class members received. DuPont agreed that they would support and pay for a separate award by the court of fees not to exceed 25.5 percent of the settlement's total value. When the $10 million estimated water-treatment costs and $5 million estimated health-study costs were added to the $70 million DuPont was agreeing to pay in cash to the class, the total estimated value of the settlement was $85 million. At 25.5 percent, the fees to be split per our original agreement among all the law firms involved would total just under $22 million. Additionally, DuPont had agreed to support expense reimbursement of $1 million. (Our actual expenses were quite a bit more.) These amounts were approved by the court. I know it sounds like an enormous amount, and it is, but considering that more than three years of work were involved and the fees were split among multiple firms, it is not quite as much as it seems. It was enough, however, that Taft would recognize it by making me the first partner in the history of the firm to receive a bonus.

The two sides spent most of the second day together in one of the larger conference rooms, hashing out the details. After such a long, contentious legal battle that had on occasion become uncomfortably personal, with attorneys on both sides having moments when they felt their motives had been questioned and their ethics unfairly attacked, the face-to-face resolution was intense but wholly professional. The most stressful and emotionally difficult part had been accomplished. We were now legal technicians assembling a complex puzzle, the pieces of which had already been formed. We only had to make it all fit together. Around 6:00 p.m. we felt the picture was finally complete. There was no sense of comradery. We would never mistake the relief we all felt for budding friendship. We shook hands cordially and walked out of the room after two very long days with a settlement in principle. There would be a few more hurdles and judicial steps to get over before the agreement was inked and finalized, but it was a major milestone. I was happy that we had been able to negotiate a deal that addressed the concerns that had led Joe and Darlene Kiger to me in the first place. They weren't after money; they wanted to know how much risk they really faced, and they wanted clean water and medical monitoring. They were getting what they wanted. And if Joe and Darlene were happy, the class members they represented would be happy, too.

. . .

I wish I could say we all danced on the tables, dizzy with victory. In reality, we were mostly just exhausted and relieved and ready, finally, to relax. That phase didn't even survive our team's quiet celebratory dinner at one of Boston's downtown seafood restaurants, whose name I can't even remember. What I do remember vividly is that before the plates were cleared, we all recognized that in order for this whole plan to work, many things would have to come together. The Science Panel would need access to an unprecedented volume of good community data related to PFOA exposures and individuals' medical histories. As DuPont undoubtedly knew, since Dahlgren's study had been viewed as inadequate, we would need more. New data would have to be collected from our class, and it would need to be bulletproof, both to withstand DuPont's negative scrutiny and to pass muster with the Science Panel. Most of all, there needed to be enough of it.

This last part was the big challenge.

We needed a rigorous study of the entire class on a scale that, as far as we could determine, had never before been attempted. We needed a way to document the individual exposure of each of tens of thousands of people, collect data on how much they might have been exposed to over time, and correlate that with their entire medical history. This was where the uniquely biopersistent and bioaccumulative nature of PFOA in human blood actually provided a one-of-a-kind opportunity. For every day a class member had been exposed to contaminated drinking water, there would likely be more PFOA building up in his or her blood. If we could get blood samples, they could confirm not only current exposures but possibly even past exposure levels, including past water concentrations.

What quickly became clear was that we needed blood. Lots of blood. Which meant we needed people who were willing to get stuck with a needle to draw it. It had been a challenge for Dr. Dahlgren to get six hundred people to show up and fill out a survey. How would we ever get tens of thousands of volunteers to give blood? And even if we could pull that off, we would also need accurate health histories. Most people remember seeing the doctor about a major illness. But inquire about dates and diagnoses and be prepared to get information of questionable accuracy.

We would have to jump through a lot of hoops, meet the extraordinarily high standards for epidemiological data, and make sure no errors were made that could compromise the integrity of the data. We knew that DuPont would put all their resources into attacking any results confirming that low-level chronic exposure to PFOA caused disease. This was the whole ball game. One mistake, however accidental or incidental, could jeopardize the study.

If that sounds like an impossible task, it almost was. I'm pretty sure DuPont was thinking so, too. The odds of getting the volume of data necessary to confirm "probable links" in a community of seventy thousand people were roulette-wheel odds or worse. The company was counting on our failure.

What they weren't counting on was our Big Idea.

• • •

The Big Idea arose at that postsettlement seafood dinner and was refined over the following few weeks. Our inspiration for it boiled down to an aphorism: Money can't buy health or love, but it can buy motivation.

As the lawyers for the class, we were the ones who would now have to explain to the court and the class members how the $70 million cash payment from DuPont would be handled. We would have to demonstrate how our plan would benefit the class. The judge would have to approve both the plan and the way we notified the class about the plan. DuPont would have no say in how we handled the money.

As per the settlement, we were expected to cut $50 million in checks to the class and use the remaining $20 million for "health and education projects." But if every single one of our seventy thousand class members did the paperwork to get his or her share of the settlement, each one would get a check for a little less than $715. That might be enough to buy a decent flat-screen TV, but it wasn't going to cut it as compensation for a lifetime of elevated cancer risk.

And it wouldn't ensure that the proper studies were actually done to answer their questions about PFOA.

I doubt the waitstaff had cleared the appetizer dishes at our celebration dinner before Harry Deitzler confessed that something was nagging at him: How could he face his friends and neighbors in Parkersburg if he walked away from this with a nice attorney's fee and paid his neighbors in the class just a few hundred bucks apiece?

"We'll look like money-grubbing lawyers," he said. "We have to do better."

I agreed completely.

That's when the Big Idea materialized: Why not use the entire $70 million to pay class members to participate in the community study?

The financial incentive was the only conceivable way we could get the astronomical numbers of participants needed to supply enough data for the Science Panel to be able to confirm probable links to specific diseases. Probable links were what we needed to trigger future relief for our class in the form of continued clean water, medical monitoring, and potential personal-injury recoveries.

Everyone would win. (Except DuPont.)

But then the tricky question loomed: How much would it cost? How much would we need to pay people to participate in the numbers needed to pull this off?

Harry thought of the focus groups his law firm often conducted. If it sent out twenty letters offering a $75 incentive to join a focus group, they could expect about two participants. If it offered $100, they might get six to ten. But we didn't need six to ten. We needed tens of thousands.

He subtracted the costs he thought we'd need to spend on processing the data. Then he divided up the remainder among the class members and found a number he thought would work. (I was happy that someone else was handling the math problems.) He imagined a family of four coming in with the kids and walking out with four times that number.

"Four hundred bucks," he said. "We need to pay people four hundred bucks."

That meant a family of four would walk away with $1,600. More important, it would be a huge incentive for people to participate in the health study. And getting as many people as possible to participate would make it more likely that class members would eventually receive something far more valuable than a onetime cash payment: ongoing medical monitoring for early detection of disease or illness, and the right to sue individually should they become ill with a linked disease.

"Why four hundred bucks?" someone asked.

"For four hundred bucks," he said, "*I'd* get stuck in the arm!"

I loved it. So did the rest of the group. Would some people prefer to get a simple, small check with no strings attached and call the whole thing over? Maybe. But we thought that most people would opt for the ability to get real answers about PFOA—plus cash.

So instead of using only $20 million of the $70 million in cash from DuPont for a health and education study, we would use essentially the entire $70 million for the blood-collection project and study. After all, the $70 million had been designated "for class benefit." The study would benefit *the entire class*—even those who opted out. Everyone drinking the water would benefit from the knowledge of what the chemicals in it did. The study would advance the science, guide public-health measures, and inform future studies. Its benefits could reach well beyond the class and be a scientific baseline for years to come.

We were excited and proud of ourselves for coming up with such a creative solution, but we knew the hard part would come in the weeks and months ahead as we tried to make our bold plan a reality. We got a quick sense of how tough it would be when we sat down to sketch out a methodology for the study to give the Science Panel. Once again, we couldn't find a model. All the experts we asked told us they'd never heard of anything like it being done before, anywhere. We'd have to build it ourselves.

Act III

THE WORLD

27

THE STUDY

September 2004
Parkersburg, West Virginia

H arry, Ed, Larry, and I sat on one side of the table in a tiny kitchen, prepared to make an audacious pitch.

On the other side of the table sat Dr. Paul Brooks, who, three decades back, had built the house where we were gathered, and Dr. Art Maher, Brooks's longtime professional rival. The two men had recently retired from running Parkersburg's two hospitals. As the directors of a small town's hospitals, they had been among the most respected medical authorities in Parkersburg.

Given the almost absurd difficulty of the mission we wanted them to take on, not to mention its intrusiveness into the lives of their neighbors, they would need to trade on every ounce of that hard-won respect.

Harry Deitzler had suggested the two of them. He had known the men since he'd been elected as Parkersburg's prosecuting attorney in the 1960s. Both Brooks and Maher were experts in the kind of collection of medical records and testing information the study would require. And, now retired, they would have no conflicts of interest, no motives other than producing a thorough and reliable study. Harry assured us that these two men would be just the dream team we needed.

"If we can get both hospitals," Harry had said, "nobody can doubt the credibility of this thing."

I had never met these men before, but I instantly felt a connection with them. It was one of the few times in my life I have entered a group of strangers and felt immediately as though I was with friends. They were highly intelligent, accomplished businessmen and leaders in their community, but they put on zero airs.

Brooks, a stout man in his seventies, spoke with a gravelly voice as he greeted everyone with an inexhaustible good humor. I knew that voice. It was the voice of my childhood. I felt as though I were listening to Grammer's brothers telling stories at my great-grandparents' house, which still stood just a few blocks away from the house where we now sat. His West Virginia cadence, the soft southern drawl, the way a trip to the grocery story could turn into a yarn—I liked him instantly.

One look at Art Maher set me back in my chair. If Brooks sounded like my great-granddad, Maher looked just like him: the same physique, bald head, and wire-rimmed glasses. That was my "granddaddy," as we had called him, the one who had died of prostate cancer when I was twelve, and I still missed him, so it was strangely comforting to be sitting across from his doppelgänger.

Brooks had been the vice president of medical affairs and operations at Camden Clark, a hospital near the river in downtown Parkersburg. It was a hospital I remembered well from my childhood. My dad had had to go there one night to have a plug from his transistor radio removed from his ear. On another visit, we had made a late-night ER run to have the wheel from one of my Matchbox cars extracted from my nose. (Don't ask.)

Maher, with an advanced degree in public health, had been the CEO of St. Joseph's, a 350-bed Catholic hospital down the street from Camden Clark. Founded in 1900 by West Virginia's first Roman Catholic bishop, the hospital was run by the Sisters of St. Joseph, who accepted many poor patients as charity cases. If the nuns trusted the guy, I figured we could, too.

Brooks and Maher had never worked together. As administrators of competing hospitals, they knew and respected each other, albeit from a distance. As we laid out the facts and the history of the whole PFOA saga, Brooks and Maher were taken aback. Despite the local coverage of the water findings, there had been zero discussion of PFOA in any of the medical journals that both men tended to read. Neither had any awareness of the decades of prior company research into the medical risks related to PFOA. They were appalled. How had they treated patients in the shadows of a plant that had known about this—for decades? How many lives could have been saved if the doctors had known? They were incredulous.

I laid out the broad brushstrokes of the scientific story I'd learned through my years of digging. As he took it in, Brooks shook his head in amazement.

"There's just too much smoke," he said, "for no fire."

Once they understood the problem, we laid out our proposed solution: a study that would settle the question once and for all. Their task would be to collect the raw data, plentiful and pristine, that our Science Panel would require. Brooks and Maher hung on every word as we explained the scope of a project that was quite possibly the most ambitious community epidemiological study ever undertaken.

I knew we were asking for the moon. We needed not only blood samples but medical histories from as many class members as possible. A self-completed survey wouldn't cut it. Every self-reported illness would have to be cross-checked with doctors and verified by medical records. As anyone who has ever navigated the health care system knows, that's a task well north of heroic and just south of impossible.

"How many people do you think are in the class?" Brooks asked.

"Around seventy thousand people," I said. "How many do you think you can do?"

I was secretly gunning for somewhere between fifteen thousand and twenty thousand, which is the number my experts suggested would be safe for finding statistical significance when looking at rare forms of cancer. I hoped that was not too ambitious.

"Sixty thousand," Brooks said.

My jaw dropped.

"If we can't do seventy-five percent," he said, "we ought not to even take the job."

If they could pull that off, they'd be even better than Harry had said. But they still hadn't accepted.

The two men turned to Harry—the man they knew the best in the room—and asked, "What results are you trying to get out of this?"

Harry looked at both of them, looked at me and Ed, and made our position crystal clear: "We don't care, as long as it's the truth."

"On that condition, we'll do it."

"We'll let the cards hit the table where they hit," Brooks said.

Those were our guys, no doubt about it.

• • •

Even with the settlement complete and the study under way, we were quickly reminded that our trials, both figurative and literal, were not over.

We had agreed with DuPont to issue a joint press release immediately after we had signed the agreement on September 4. The release had been picked up nationally by *USA Today* and by the local West Virginia and Ohio papers. DuPont had published the press release on their corporate website, and we had filed a preliminary summary of the deal with the court, along with a plan for formal notification of all class members. We would publish the formal class notice in various papers and send it by direct mail to addresses obtained from the affected water districts. The notice informed all class members that they had a right to opt out of the class, for whatever reason, or to object to the settlement at a final hearing in February 2005. It also explained that, in addition to the payment of up to $400 for the questionnaire and blood draw, they would receive free blood tests for a myriad of endpoints (cholesterol levels, liver enzymes, etc.) and not only PFOA but PFOS and other C5 to C12 compounds, valued at an additional $500. Preparing all these notices amounted to a pretty substantial effort on our part, as well as substantial costs for the required advertisements and mailing, paid for by DuPont.

As we were otherwise occupied, DuPont continued to seed the

scientific literature with more data to support their claims. Just as residents of the Mid–Ohio Valley were learning of our science-based settlement, DuPont announced the results of their latest study of their workers. No surprises there: DuPont claimed that a $1 million Washington Works PFOA worker study had found "no adverse health effects," aside from a minor increase in cholesterol levels.

That same day, EPA aired news of its own on PFOA. "We have identified potential for carcinogenicity," said Charlie Auer.

This was a stunning admission. If EPA, which had seemed willingly blind to PFOA health risks for years, was finally coming around, my optimism surged that our soon-to-be-handpicked science panel would follow suit. This wasn't the common cold they were talking about; it was cancer, the Big C, which had a way of grabbing people's attention. The agency's newly revised PFOA risk assessment reported a potential risk of developmental and other adverse effects as well. I allowed myself to feel some pride of vindication. That may have been the last time I was elated about anything the agency said about PFOA, and I probably should have known better even then. There were warning signs. The carcinogenicity announcement came with a caveat, a "but" clause: "But the information is not sufficient for conclusions to be made." And the other risk findings were said to need further peer review. EPA would consult its own Science Advisory Board, separate from our Science Panel, expressly for that purpose.

EPA's tepid announcement, with its mixed signals and ambivalent conclusions, barely moved the market to sneeze. The DuPont stock price plunged all of one nickel.

DuPont then put out yet another press release: "Comprehensive Scientific Study Confirms Consumer Articles with DuPont Materials Are Safe for Consumer Use." And who had helped orchestrate this new "scientific" study? Consultants at ENVIRON—a private consulting firm that had been hired by DuPont as part of their legal defense team.

It seemed that DuPont was trying to build a firewall between the everyday use of their most successful consumer products and the manufacturing process used to make them. In the very same press release,

DuPont promised to reduce nationwide emissions of PFOA by more than 98 percent by the end of 2006. Certain products would be reformulated to curb emissions down the supply chain. To top all that off, DuPont was working on "critical technologies" to reduce emissions in the manufacturing process, and they would actually share those innovations royalty free with other fluoroproducts companies.

I felt a surge of optimism: this was the victory I had been working toward for over half a decade, finally in sight. DuPont was making a strategic retreat, but at double-step pace. I was further buoyed when, just as DuPont was getting rolling with their PR offensive, the US Department of Justice stuck a stick through the spokes in the form of a subpoena. The DOJ's Environmental Crimes Section subpoenaed DuPont on behalf of a grand jury seeking records on PFOA. Headline news of a criminal investigation caused a nationwide stir.

The market's response? The DuPont share price dropped a measly eighty-three cents.

Did investors care at all? Some did. A shareholder group called DuPont Shareholders for Fair Value asked the Securities and Exchange Commission to investigate DuPont's handling of PFOA disclosures—to shareholders. The disclosures in question were not about the human health risks of the chemical but about the financial risks of PFOA litigation to their investment in DuPont stock.

In subsequent SEC filings, DuPont made the financial risks crystal clear: a complete ban or restriction on PFOA could impact $1 billion in DuPont sales.

· · ·

DuPont wasn't the only company in trouble for PFOA and its chemical relations—known collectively as perfluorochemicals (PFCs) or per- and polyfluoroalkyl substances (PFAS).

At a 3M plant in Minnesota, the news wasn't just about PFAS contamination; it was about the alleged efforts to conceal it, another cover-up attempting to bury the science. The whistle-blower, Fardin Oliaei, had a

PhD in environmental sciences and was employed at the Minnesota Pollution Control Agency (MPCA). Her research had found unprecedentedly high levels of PFAS in fish in a stretch of the Mississippi River that flowed past the Minnesota plant where 3M had made the PFOA it had sold to DuPont for more than fifty years. In an official whistle-blower complaint, she alleged that she had endured threats and reprimands from her own state government agency, including from the head of the agency, Sheryl Corrigan, a charge Corrigan denied.

By now I wasn't shocked to learn where Sheryl Corrigan had worked before overseeing the state's regulatory agency: between 1996 and 2002, she had worked for 3M as a high-level corporate environmental manager.

Ultimately, Oliaei agreed to drop her lawsuit and resign in exchange for $325,000. "It is a sad commentary on the state of affairs in Minnesota when the state government will shell out big money just to keep its scientists from doing research," said a spokesperson for a whistleblower-protection organization.

• • •

Even for companies such as DuPont and 3M, it was getting harder and harder to suppress the alarms being sounded all over the country by civic groups, environmental activists, and still more scientists.

Around Parkersburg, Jim and Della Tennant had rallied hundreds of local residents to sign a new petition objecting to the further renewal of the Dry Run Landfill permit. Now that the news of the government's criminal investigation of DuPont had been widely reported, the die-hard support for DuPont had soured, in many cases, into smoldering anger.

Despite the success of the petition drive, the West Virginia DEP renewed the landfill permit anyway. Then, just months after the permit was renewed, DuPont disclosed two leaks from Dry Run Landfill and PFOA levels of 151 ppb measured in Dry Run Creek—a level even higher than the ridiculous 150 ppb CAT Team screening level from 2002. And in the same creek where all this had begun.

Across the river in Little Hocking, residents were boiling over the findings of a new federally funded community blood study unrelated to our still nascent health study. Since DuPont had quashed our injunction to force them to let us use their blood-testing lab, PFOA testing methods had slowly started to be adopted by other labs, enabling University of Pennsylvania researcher Ted Emmett to test the blood of hundreds of customers drinking the Little Hocking Water Association's contaminated water. According to his tests, their PFOA blood levels ranged from 298 to 360 ppb—levels "sixty to eighty times higher than the general population."

Elsewhere in Ohio, a citizen activist group was sending letters to major fast-food chains asking them to confirm whether their fast-food wrappers or pizza boxes contained PFAS. Some of the country's best known fast-food and packaged-food brands were now on the hot seat.

Things were happening in the reddest of red states and the bluest of blue. In Alabama, residents were suing 3M over PFAS contamination of soil and groundwater in Decatur. The litigation is still pending. In California, the state legislature was passing the first state bill to set up a biomonitoring program to test residents' blood for various chemicals.

A former DuPont employee joined the chorus. Glenn Evers, a retired chemical researcher for DuPont, spoke to the press, claiming he had warned the company years before about the potential for PFOA to migrate from food packaging into the human body, only to find himself "pushed out" by the company.

DuPont refuted his allegation, saying that all such products were approved by the FDA. "These products are safe for consumer use. FDA has approved these materials for consumer use since the 1960s, and DuPont has always complied with all FDA regulations and standards regarding these products."

That response could not drown out Evers's warning, which echoed across the AP wire: "You don't see it. You don't feel it. You can't taste it. But when you open that bag, and you start dipping your French fries in there, you are extracting fluorochemical and you're eating it," Evers said. "DuPont thinks they have pollution rights to the blood of every American—every man, woman, and child in the United States."

• • •

While Paul Brooks and Art Maher began preparing a plan for the PFAS blood and raw-data collection effort funded with the $70 million cash settlement, we worked with DuPont to choose the three independent epidemiologists to serve on our Science Panel. Since they were the ones who would make use of all the data Brooks and Maher were collecting, the composition of the Science Panel would be one of the most critical decisions in our litigation.

Each party prepared a list of around two dozen candidate epidemiologists. We looked for candidates whose research history was strictly academic, with no prior work in litigation of any kind. If they had ever served as an expert witness or been hired as a consultant for a case, they were presumptively disqualified.

I spent countless hours researching potential candidates and reviewing their backgrounds, résumés, and publications. I sought guidance from our experts on the reputation and standing of the candidates that might not be apparent from their public materials. I searched for every piece of information we could find that might give us any insight into potential biases. Their funding sources had to be identified and examined. I had learned from this whole experience that we had to be thorough—and extremely careful. There were a lot of prominent people who were vetoed. But we finally built a list of highly accomplished candidates, as did DuPont.

We then exchanged lists and researched each other's nominees. Either party could strike any candidate from the list for any reason. That narrowed the pool considerably. Once we had winnowed it down to a group of nonvetoed candidates, we culled the list further through phone interviews. A number of them were either not interested or too busy to take on the considerable time commitment we required.

The final step was conducting face-to-face interviews. When we learned that DuPont's lead counsel in the class action, Larry Janssen, would be handling the interviews for DuPont, I thought it best that someone from our team other than me accompany him.

In all that time, Janssen still seemed to actively avoid eye contact with me, and when he spoke with our team, I felt that he typically directed his comments to others. Ever since we had revealed our plan to use DuPont's $70 million payment under the class settlement to fund the PFAS blood and health data-collection program, I had sensed even greater tension in my communications with him. Maybe I was just imagining it all, but I didn't want any potential personality clash to interfere with the interviews. They were too important. I called a meeting of our legal team, and we agreed that someone else would have to go.

Larry Winter stepped up. "I guess I drew the short straw," he would joke. For the next five months, Winter and Janssen would have to work together closely, negotiating a travel schedule and hitting the road together to interview the top candidates.

Taking intermittent trips together, Winter and Janssen worked their way down the final list, collaboratively interviewing each candidate and comparing notes on the drive to the airport. At the end of the day, Winter would circulate his notes to our legal team, and I would follow up on any needed research and further investigation of the candidates.

By February 2005, we had a mutually approved Science Panel.

Dr. Tony Fletcher, PhD, worked for the London School of Hygiene & Tropical Medicine, one of the most respected public-health research institutions in Great Britain. Dr. Fletcher had worked in environmental and occupational epidemiology for twenty-five years, evaluating evidence of human health risks and studying water and air pollution throughout Europe, and had served as president of the International Society for Environmental Epidemiology.

Dr. David Savitz, PhD, was the Charles W. Bluhdorn Professor of Community and Preventive Medicine at the Mount Sinai School of Medicine.

Dr. Kyle Steenland, PhD, was a professor in the Departments of Environmental Health and Epidemiology at Emory University's School of Public Health and had spent twenty years at the National Institute for Occupational Safety and Health in Cincinnati.

The future of our seventy thousand class members, and possibly the ultimate fate of PFOA, was in their hands.

. . .

July 2005
Parkersburg, West Virginia

If there was one thing that lit a fire under Paul Brooks, it was being underestimated. He knew some guys in fancy suits believed the little team of West Virginians he had recruited to generate big data on the links between chronic drinking water exposure to PFOA and disease could never pull this thing off. After all, he was literally asking people to bleed for him.

"You'll be lucky to get ten thousand people to fill a survey out online," someone said.

"Fifteen" was another guess. "*Maybe* twenty thousand."

Brooks looked forward to the day they would realize how very wrong they had been.

He had told us his outrageous goal: sixty thousand. Secretly, he was aiming higher.

Our proposed plan to use DuPont's $70 million payment to pay class members to provide blood and health data—$150 if they completed a questionnaire and another $250 if they provided a blood sample—had been formally approved by the class members and the court during the hearing in February 2005. That approval had given Dr. Brooks and Art Maher the green light to move forward with their project to collect that data—a project they called "The C8 Health Project." They combined their last names to form the name of the new company that would implement the project: Brookmar, Inc.

The technical challenges had been so intense that it had taken five months of innovation by a handpicked group of IT geniuses to get to this point. Brookmar flipped the switch on an Internet collection system the IT wizards had built from scratch and held its collective breath. It was like turning the ignition of an experimental rocket engine that could blast off or might just explode. From the control screen they watched as the first person entered the system. Then another and another and a dozen more began filling out the survey in a flow that swelled and wouldn't stop.

By the end of July, the nurses were drawing blood. People marched out of the trailers the team had set up in multiple locations across the community with a Band-Aid and a check. Word spread like a brush fire on a windy day that the promises were real. If you showed up to a testing site, you'd fill out your questionnaire, have your blood drawn, and walk out within the hour, check in hand. I had a feeling it wasn't just about the money.

By the end of August, after just two months of operation, nearly thirty-five thousand people had filled out the seventy-nine-page questionnaire—it took about forty-five minutes—and the vast majority of those, eager for the $250 additional perk, had also given blood.

The obstacles—transportation, illiteracy, Internet access—turned out to be very surmountable. People helped one another. Tech-savvy kids helped grandparents and neighbors with limited Internet skills. Many people filled out a hard copy and asked a family member to transcribe it online. Churches and libraries helped organize and mobilize.

Our plan was actually working.

28

THE SECOND WAVE

August 2005
West Virginia, New Jersey, and Minnesota

A s the C8 Health Project was drawing blood, the debate over PFOA was drawing new scientists out of the woodwork. Our litigation had triggered an explosion of scientific interest in PFOA, and new research projects were beginning all over the world—studies that had no connection to our case or to DuPont. Our settlement had granted the Science Panel full discretion to consider relevant data from any of those studies, not just DuPont's or 3M's own self-serving studies or data collected through the C8 Health Project.

Adding to my sense of growing momentum, in June 2005, EPA's Science Advisory Board—the agency's mechanism for peer review—looked at the latest draft of EPA's risk assessment. Based on the data its members had seen, which was essentially all the data I had been sending, they recommended that EPA revise its risk assessment and upgrade PFOA from a "suggestive" to a "likely" human carcinogen. This was an important technical distinction that could impact regulatory requirements down the road. It also had another, more immediate impact: headlines suggested that EPA had earlier downplayed the cancer risk and its own advisory board was now calling it out.

While the science was mounting against PFOA, however, a controversial faction of the scientific community was publicly defending it. The American Council on Science and Health—a nonprofit "consumer education consortium" of scientists—was vocally saying that there was "not a shred of evidence" that either Teflon or PFOA posed any cancer risk whatsoever. It claimed that "junk science" was being used to set public policy, and it criticized EPA for trying to protect us from "cancer risks that do not exist."

Scientists often disagree on a lot of things, but that group was a great example of why it's important to trace the money behind the science. The innocuous-sounding American Council on Science and Health was founded in 1978 by a Harvard-trained public-health scientist as a counterpoint to advocacy groups such as the Environmental Working Group, which it accused of fearmongering with claims it said lacked scientific basis. The Council billed itself as an independent group of scientists aiming to debunk the bogus scientific findings it claimed were driving political decisions in public-health and environmental policy.

But critics called the Council an "industry front group" funded by corporate donors it refused to disclose. Leaked internal documents showed a donor lineup of household names in industries ranging from Big Tobacco to pharmaceuticals to petroleum and, yes, chemicals, including 3M. They were the same industries the group fiercely defended in the science wars that played out in the media.

It was something to keep in mind when hearing that group defend Teflon as "the poster child of modern technology." Teflon "made our lives easier and more enjoyable," the council maintained, and it was Teflon's very success that had made it such a "ripe target for those who spew chemical-phobia in their crusade to eliminate the tools modern industrial chemistry has given us."

At the same time, it seemed that the publicity surrounding our class settlement in West Virginia and the increasing media attention to the health risks of PFOA were attracting new lawyers and new lawsuits. In July, attorneys in Miami, Florida, announced they had filed a $5 billion class action lawsuit against DuPont over Teflon. The complaint? Failing to warn consumers about the dangers of fumes emitted by their nonstick cookware, which, they claimed, included PFOA. Tests run by consumer advocates allegedly showed that Teflon-coated pans heated above 570 degrees gave off toxic fumes, the same kind of thing that had been shown on the *20/20* segment back in 2003 when the folks there had fried bacon. The lawyers said that DuPont had sold $40 billion worth of coated cookware over forty years, and their class of plaintiffs could include nearly every American who had ever purchased a pot or pan coated with Teflon.

I was not focused on the risk to consumers of using Teflon products, and the consensus of those who were would eventually settle on the idea that drinking tap water contaminated by manufacturing waste posed a far bigger threat than frying eggs for breakfast in a Teflon pan. But the fact that others finally seemed to be recognizing the seriousness and scope of the PFOA problem encouraged me. The Department of Justice's criminal investigation was still moving forward. (Its representatives even showed up at Taft with a subpoena and spent four days digging through my PFOA files.) Legislators in certain states were beginning to focus on PFOA and had started looking at possible legislative or regulatory responses. Canada was considering banning not only PFOA but the entire family of PFAS.

Our legal team was getting its own recognition. During all my prior years of defense work, I had never won an award, so the irony wasn't lost on me when my first big recognition came from the Trial Lawyers for Public Justice, which honored our entire DuPont legal team with its 2005 "Trial Lawyer of the Year Award" at a ceremony in Toronto.

The year closed with a big announcement from EPA and DuPont. For failing to report the early PFOA toxicity studies and drinking water contamination data to EPA, DuPont would pay a $10.25 million penalty and provide another $6.25 million worth of "environmental projects." The $16.5 million total was a small fraction of the rumored $300 million in potential fines that had swirled in the media, but it was still claimed by EPA to be "the largest civil administrative penalty the EPA has ever obtained under any environmental statute." Regardless, PFOA was still not a regulated chemical. The fine was for failure to report studies that could, under the right circumstances, be the starting point of the very long road to official regulation.

January 2006 kicked off with more encouraging news. EPA announced a new PFOA Stewardship Program, calling on makers of PFOA to slash emissions and reduce the amount of the chemical used in products. The goals were ambitious: a reduction of 95 percent (of year-2000 levels) by 2010 and complete elimination by 2015.

I was overwhelmed. If EPA stuck to its word, it would mean that the manufacturing of PFOA would be eliminated in the United States entirely. I couldn't stop thinking about what a long way we'd come from the Cattle Team report six years earlier that hadn't even recognized PFOA's existence.

DuPont issued statements trumpeting their commitment to the program, while still continuing to use the careful language that did not acknowledge any real problems. And (with the new Teflon pan litigation pending) one of their press releases also noted that EPA had "consistently said there is no information that would indicate any concern" over consumer use of household products containing the chemical.

EPA even claimed to be working toward adding PFOA to the list of chemicals that had to be reported under the Toxic Release Inventory Program, which would have been the big win, finally making PFOA a "listed and regulated" substance, moving closer toward triggering Superfund cleanup responsibilities. As with all EPA-generated optimism, I would painfully discover, that hope would never be fulfilled.

When reporting on the new PFOA phaseout program, ABC News referred to Sue and Bucky Bailey. "Today they called the action long overdue." Sue was interviewed by another TV station and expressed what the milestone meant to her: "I feel really victorious today."

A few months later, EPA even publicly rejected the ridiculous West Virginia 150 ppb CAT number and announced its own new consent agreement with DuPont. The new agreement required DuPont to provide clean drinking water if PFOA was found in drinking water above 0.5 ppb in communities impacted by the Washington Works plant—a minuscule fraction of the CAT Team's number for "safe" drinking water. Because

PFOA was still unregulated, EPA could not force DuPont to use the much lower 0.05 ppb threshold for providing water treatment that we had gotten DuPont to agree to use under our class settlement—or any other number. The consent order reflected only what EPA could get DuPont to agree to. Nevertheless, getting DuPont to move from 150 ppb to 0.5 ppb—"voluntarily"—was huge. Plus, it affirmed my earlier suspicions that even DuPont's own scientists had known that the 150 ppb number was absurd.

For the first time in my life, I felt as though I was truly making a difference. In the regulatory world. In the scientific world. In the legal world.

And, most important, in the physical world. To people like Sue and Bucky.

All the stress and anxiety I had been battling for seven years since taking on those contingency cases—and watching the unbilled hours and expenses escalate—finally evaporated.

I felt what I never imagined I'd feel: a plaintiff lawyer's sense of victory. I had finally secured relief for my clients: the Tennant family, as well as Joe Kiger and his community. The rest of the world was finally seeing PFOA as I had been seeing it. It was now clear that the risk I had asked my firm to take on had been justified.

And for the first time, I felt that I had proven myself. I was no longer feeling insecure and anxious about whether my work was a fit for the firm. On the contrary, I was actually getting firmwide recognition and accolades. I had just brought in the largest fee in the firm's history.

My partners even gave me a "trophy" of sorts. During a partner meeting, I was called up to the front of the room and presented with a bright red Teflon frying pan engraved with these words:

TO ROB BILOTT
FROM HIS PARTNERS AT
TAFT STETTINIUS & HOLLISTER
ON THE OCCASION OF THE
DUPONT TEFLON SETTLEMENT
JUNE 30, 2005

I detest being the center of attention. (I even dread birthday parties where I have to open presents with everyone watching.) But on that day I was truly proud. And happy. At last, all those "dark horse" years at the firm were finally behind me.

With the active litigation part of the case finally settled, the Science Panel in place, and the C8 Health Project in full swing, I even had some time to relax and spend some quality time with my family. We splurged on a weeklong Disney Cruise in the Caribbean. The boys, now seven, five, and almost four, had a ball. Sarah and I got some much-needed time to recharge and reconnect. That summer was one of the happiest times in my life.

Along with my new confidence and new feeling of security came more new clients. As the details of West Virginia's contamination problems began to circulate, people in other states began to wonder about the water in their own neck of the woods.

Like Dr. Oliaei, who'd blown the whistle on PFOS contamination in Minnesota and resigned, other officials in Minnesota, 3M's home state, logically wondered about the 3M plant that had made and sold PFOA there for half a century. Sure enough, further state testing confirmed PFOA (and PFOS) in drinking water near 3M's plant in Cottage Grove. I got a call asking whether our "West Virginia team" could help. I soon found myself as counsel in another contingency-fee case. It was only the first. In March 2006, the Delaware Riverkeeper Network, a local environmental group, found elevated levels of PFOA in drinking water near DuPont's Chambers Works plant in New Jersey. Again I was asked to help. About a month later, new sampling showed PFOA levels in the City of Parkersburg's drinking water now exceeding, for the first time, the 0.05 ppb threshold for establishing membership in our existing class. But because the levels were below the higher 0.5 ppb level DuPont had agreed to in the new EPA consent agreement, DuPont refused to provide clean water to Parkersburg. When that community reached out for help, we agreed to take the case. We even accepted an invitation to join the team of lawyers handling the Teflon pan litigation to help advise on the science issues surrounding PFOA. All of these new cases served a very important additional function: through the newly filed cases and new discovery, I would now have the ability to continue to collect documents and talk to witnesses from DuPont—and now also 3M—and monitor what these companies were up to on PFOA issues while our Science Panel was working, even though our original class action for Joe Kiger and his neighbors was settled. I knew this was critical, given how much I had learned about PFOA that was happening outside the public's view. I went into those cases buoyed by a confidence I had never felt before.

Now I had another chance to put what I'd learned to use. I saw the path to resolving the problem for all of these communities.

It all seemed so clear.

29

DARK SCIENCE

2005–2006
Cincinnati, Ohio

After the big announcement of the PFOA Stewardship Program—its challenge to all PFOA users and producers to cut emissions and steadily reduce the amount of the chemical used in products—I wasn't the only one rejoicing. A great eruption of self-congratulation and mutual back patting commenced.

"This is the right thing to do," said EPA's Susan Hazen. "It is the right thing to do now for the environment, for public health . . . and for the agency."

EPA trumpeted the ten-year plan for virtual elimination of PFOA—with no required action under federal laws—as a glorious example of corporate cooperation without regulation. The agency touted the "100 percent participation and commitment" of the eight companies asked to participate and praised them "for exemplifying global environmental leadership."

The Environmental Working Group, meanwhile, publicly commended EPA "for their leadership."

This was remarkable progress on the heels of years of resistance. It seemed almost too good to be true. The companies were voluntarily on board. EPA was taking action. And even the lead environmental watch-dog group on the issue appeared satisfied. The momentum seemed un-stoppable.

Then, all too abruptly, it stopped.

After the PFOA Stewardship Program fanfare, EPA's engagement in all PFOA matters essentially came to a screeching halt. Its plans to add PFOA to the toxics release inventory? Never happened. The risk assessment initiated in 2003? Never finalized. The complaint against DuPont? Settled. The Department of Justice criminal investigation? Terminated.

EPA, in fact, would take no significant additional action on PFOA for another decade.

Why would EPA back off so suddenly and completely? It was obvious. Or it should have been. The whole relationship between EPA and DuPont was far too cozy. I'd gotten that uncomfortable sense many times over the years, most recently when I'd attended a public meeting of EPA's Science Advisory Board. I had been taken aback when the EPA employee leading most of EPA's risk assessment work on PFOA had walked into the meeting hall elbow to elbow with the head toxicologists on PFOA for DuPont and 3M. They had been laughing and talking like old friends.

I didn't give too much weight to it at the time— it could have been perfectly innocent. After all, this was the federal EPA—not some small state's agency, such as the West Virginia DEP or even the Minnesota PCA, that could easily be co-opted by big local employers. But now, thinking back, something about the memory of that EPA advisory board meeting in 2005 nagged at me. I wondered if maybe the federal EPA—and the whole federal review process—also wasn't so independent after all. The documents that had continued to arrive via discovery thanks to my involvement in the new cases revealed a pattern I had been noticing in EPA's public statements about PFOA. Around the time the PFOA Stewardship Program had been announced, EPA had released a statement oddly reminiscent of DuPont's own refrain: "To date, EPA is not aware of any studies specifically relating current levels of PFOA exposure to human health effects." That same day, DuPont had sent out a press release quoting EPA's statement. I suspected that the timing was more than a coincidence. I believed it was part of a quid pro quo: DuPont would agree to participate in EPA's splashy phaseout initiative in exchange for EPA's issuing a statement supporting DuPont's PFOA mantra: no known human health effects.

So the next month it must have come as a nasty surprise to DuPont when EPA's Science Advisory Board reiterated its recommendation to reclassify PFOA as a "likely" human carcinogen. The widespread negative publicity that followed significantly worried consumers.

The new documents from DuPont that I was poring over revealed the depth of concern at the company. Among those documents was an email from Susan Stalnecker, the head of DuPont's "Core Team" on PFOA issues, sent directly to CEO Chad Holliday with the subject line "URGENT." Her message advocated using EPA to protect DuPont's exposed flank. She wrote, "In our opinion, the only voice that can cut through the negative stories is the voice of EPA. We need . . . EPA to quickly (like first thing tomorrow) say the following: 1. Consumer products sold under the Teflon brand are safe. . . . 2. Further, to date, there are no human health effects known to be caused by PFOA."

Two weeks later, in a conference call about the PFOA Stewardship Program, DuPont got exactly what they had sought. EPA's Susan Hazen announced, "The agency does not believe that consumers need to stop using their cookware, clothing, or other stick-resistant, stain-resistant products." She added that "the agency has no evidence that use of these products leads to PFOA exposure to the consumer."

My suspicion that DuPont and EPA had privately hammered out some kind of mutually beneficial deal prior to the PFOA Stewardship Program got a big boost when I came across a document we came to refer to as "the silver bullet memo." In it, Michael McCabe, a former EPA deputy administrator who had been hired by DuPont as a consultant, laid out an exit strategy for DuPont and EPA. "I have always believed that the way out of this quagmire," he wrote, "is to publicly roll out the process changes that essentially eliminates P***." (DuPont had started using code words for PFOA in some internal documents, presumably to evade electronic keyword searches in future discovery. Other code words: "the

P-issue" and "Project Elephant," which I took as a reference to "the elephant in the room.")

DuPont would be hailed as the "global leader" in voluntarily reducing emissions, developing alternate technology for making products without PFOA, and committing to a complete phaseout of the chemical within ten years. In return, EPA could publicly take credit for fixing the PFOA problem.

By this point, PFOA had become a major issue for EPA as well. Given the number of products the chemical had been used to make and the number of places across the country where those products were made (and likely disposed of), it was becoming increasingly clear that the chemical was likely in air and water all over the world. It had been established that the chemical was likely already in the blood of every man, woman, and child in the nation. The agency's own draft risk assessment was being interpreted by some to suggest that those levels might present health risks to sensitive populations. Some were asking how in the world EPA had allowed all this to happen. And more important, what was the agency going to do now to fix it? EPA's lawsuit and DOJ's criminal investigation had emphasized the severity and magnitude of the problem. But neither had gone anywhere or done anything to fix the actual problem.

As public and media pressure on EPA intensified, DuPont's plan would finally give the agency a way out. But as laid out in the silver bullet memo, the offer came with a very important condition: EPA would have to "create a level playing field" by requiring the entire industry to make the same commitments. To me, that explained the PFOA Stewardship Program and also DuPont's offer to share their PFOA-reducing technology— royalty free—with competing companies. In exchange for these considerable commitments, DuPont wanted some things from EPA. Another of our discovery finds from our cases springing up around the country was a series of slides from a DuPont presentation to the agency that spelled out, in no uncertain terms, "What We Need from the EPA."

First and foremost was a public commitment not to take regulatory action on PFOA pending the completion of EPA's risk assessment, which, given the speed of that process so far, DuPont had every reason to believe could take many more years.

I had to conclude that this was why EPA had backed off so suddenly and completely. (In fact, EPA would be virtually inactive on all PFOA matters until 2016.)

Moreover, DuPont wanted more than "a quiet pat on the back." (That's a direct quote from the presentation, by the way.) It wanted high-profile public credit for taking a "leadership role in making voluntary reductions." It also wanted EPA to make reassuring public statements concerning the safety of PFOA and DuPont consumer products, a request that was clearly granted.

For both sides, it was a pretty good deal. Both got a face-saving escape route from the PFOA mess. Both got to take credit for "fixing" the problem. And they could tell everyone, at last: nobody is going to use this stuff anymore. Everything is fine. End of story.

But what was the actual benefit to the public? Was anyone cleaning up the PFOA already out there? What about the PFOA in the aquifers, in the soil, on playgrounds, in backyards, all over the United States and beyond? The chemical had been spewed out into the environment for more than half a century. Given the persistent nature of the man-made chemical bond, the millions of pounds of PFOA that were already out there would likely stay in the environment for millions of years unless physically removed and destroyed. What was already out there in people's blood and bodies would also persist. The plan did nothing to address any of that.

. . .

Despite all the growing concerns and negative publicity about PFOA, DuPont's reputation in the financial world was still looking like spit-shined brass. The company was still ranking high on "best of" lists not only for their financial performance but, ironically to me at least, lists based on environmental criteria and sustainable business practices.

In 2005, DuPont was still on *Fortune* magazine's Global Most Admired Companies list, coming in at number 37. In the *Financial Times*' survey of the World's Most Respected Companies, DuPont was named number 24 in the category of "World Community Commitment." DuPont was named to the Dow Jones Sustainability Index, a financial index of the world's most sustainable companies—as it had been every year since the index had been created in 1999. Perhaps most remarkably, it was number 1 on *BusinessWeek*'s ranking of the Top Green Companies. (The list also included 3M.)

Internally, however, DuPont was churning with dissent. Many voices were now piping up, urging the company to do the right thing, and not just those of their lawyers, such as Bernie Reilly and John Bowman.

Through discovery on a new case, I found an internal memo from DuPont's Epidemiology Review Board, a committee of outside, independent scientists contracted by DuPont to provide internal guidance on epidemiological research and ethical issues. The memo pointedly struck directly at the heart of DuPont's PR strategy in surprisingly strong, even angry, language.

"Given the many gaps in understanding of population exposures to PFOA and of possible health consequences, we strongly advise against any public statements asserting that PFOA does not pose any risk to health," the ERB members wrote to DuPont in early 2006. "We also question the evidential basis of DuPont's public expression asserting, with what appears to be great confidence, that PFOA does not pose a risk to health."

Inside DuPont, that "lay it on the table" strategy hit fierce resistance. One memo to top execs urged the company to go in the opposite direction and resist greater transparency. Titled "Connect the Dots," it read to me like an outline for the company to find a way to keep government regulators, the media, the public, and, especially, trial lawyers away from the facts that could connect troubling information about PFOA. It concluded, rather plaintively, I thought, "Is there a strategy we could use to minimize the amount of information being disseminated?"

Another document that surfaced was a pitch from the Weinberg Group, a Washington, DC–based consulting firm hired by DuPont that specialized in corporate product defense. In its proposal to DuPont, it boasted of its pedigree in crisis management for corporations in challenging times. Starting with its defense of the manufacturers of Agent Orange in 1983, the letter said, the Weinberg Group had "successfully guided clients through myriad regulatory, litigation and public relations challenges posed by those whose agenda is to grossly over regulate, extract settlements from, or otherwise damage the chemical manufacturing industry."

In other words, it did glorious battle against evil bureaucrats, plaintiff's lawyers, and environmentalists.

Perfluorochemicals apparently were not new to the Weinberg Group. "[W]e already have extensive experience in helping a Fortune 40 client with a very similar compound to PFOA," the letter said. "Our experience and knowledge regarding this compound is very well established. We do not need to educate ourselves at DuPont's expense."

The letter indicated that the Weinberg Group had been following the PFOA matter—and our case—very closely. Closely enough to acknowledge exactly how much was at stake for DuPont. "[D]ue to the situation in West Virginia and the activities of the Environmental Working Group, the threat of expanded litigation and additional regulation by the EPA has become acute," the letter said. "The recent ruling by Judge Hill regarding blood testing underscores the need to act quickly and forcefully.

"DUPONT MUST SHAPE THE DEBATE AT ALL LEVELS," the letter said [all caps theirs]. "We must implement a strategy at the outset which discourages governmental agencies, the plaintiff's bar, and misguided environmental groups from pursuing this matter any further than the current risk assessment contemplated by the Environmental Protection Agency (EPA) and the matter pending in West Virginia. We strive to end this now."

In the letter, I could see my own face in the crosshairs.

The Weinberg Group's "multifaceted plan" specifically mentioned the need to "take control of the ongoing risk assessment by the EPA."

This risk assessment was the gateway to possible regulation. I assumed that was why the Weinberg Group was saying it had to be stopped.

And how did it propose to do so? With science, but the kind of science carried out under its own direction, with DuPont's interests at heart. The proposal recommended "facilitating the publication of papers and articles dispelling the alleged nexus between PFOA and [birth defects] as well as other claimed harm" and "coordinat[ing] the publishing of white papers on PFOA, junk science and the limits of medical monitoring"—in other words, deliberately seeding the scientific literature with publications that would discredit, or at least cast doubt upon, the legitimate scientific studies that indicated the real health risks of PFOA and the need for medical monitoring as a classwide legal remedy.

And if the data for such studies did not exist, the Weinberg Group could create it.

It actually suggested "constructing a study to establish not only that PFOA is safe over a range of serum concentration levels, but that it offers real health benefits."

The audacity was astounding.

30

BURDEN OF PROOF

2007–2010
Cincinnati, Ohio

While DuPont was busy muddying the waters with PR campaigns and more corporate-funded health studies, over the next three years EPA would drag its feet and fail to follow through on all the promises it had made—to the public, that is—even as new independent research made the dangers and pervasiveness of PFOA across the country and around the world ever clearer. Nonchemical calamities took over the headlines—the stock market imploded, the economy tanked, and financial structures around the world threatened to collapse—and any remaining hope I had that EPA had ever seriously intended to address the PFOA problem evaporated.

My personal world was in upheaval as well. I had thought my feelings of insecurity were behind me, but I was wrong. At work, I faced mounting expenses and unbilled time for the new cases—and new legal roadblocks were thwarting progress in those cases—while at home, family health crises rocked me to the core. Those would be the darkest years of my life. I would be tested in every way, at every turn.

Humming beneath all these events was the unrelenting tension caused by the long wait for the Science Panel's results. The three epidemiologists had essentially been given a blank check and no deadline to produce scientific results that were so rock solid that no one could ever question them—not us, not DuPont, and not their scientific peers around the world, who would all be watching. I remained convinced that the Science Panel would confirm probable links. We had seen what the separate C8 Health Project led by Dr. Brooks, Art Maher, and the entire Brookmar team could do. Completed in late 2006, their project had turned out to be one of the most successful and largest community-health-data collection programs ever undertaken—anywhere. By the time the last questionnaire was filled out and the last blood sample was analyzed, Brookmar had succeeded in gathering health and blood data from approximately sixty-nine thousand people in a community estimated at between seventy thousand and eighty thousand people. This was exactly the kind of high-quality— and massive —data required for the Science Panel to reach definitive conclusions. It was the kind of data DuPont had likely believed we could never generate. And now that we had defied their expectations, it became clear that the company was waging a war in the background, conducted through PR firms and consultants, to question and undermine the evidence.

Aware now of the power and sophistication of the forces we were up against, I realized I could not just sit back and hope everything would work out in the end. Even though the active litigation phase of the case for Joe Kiger and his fellow class members was technically settled, I was still consistently spending a massive number of hours each month (and incurring even more expenses) doing research. Virtually every morning for the next seven years, I would jump onto the PubMed federal database and scour every new publication for any potentially relevant piece of data I could find that might be of use to the Science Panel.

Sarah, ever so understanding of my perseverance, started raising her eyebrows as I would reach for my BlackBerry first thing upon waking. She was either reaching the limits of her tolerance for my distractedness or, even worse, she had accepted it as part of our lives.

Combing the Internet for scientific data was no easy task. At that point, the volume of PFOA research around the globe was exploding.

Our experts helped me interpret and understand the new data as it
emerged. But they charged for their time, and the bills added up. I also
had to monitor the studies that DuPont was sending to the Science Panel.
If those studies had been funded by DuPont or 3M (or their consultants),
I wanted to make sure the panel was aware of it. The battle wasn't over.
I knew I couldn't let up now.

. . .

My efforts had paid some dividends. Our class members, under the
settlement, were getting clean water. As soon as the settlement was approved in 2005, DuPont began funding the construction and operation
of new water-filtration plants for the public water supplies and new in-
home filtration systems for class members with private wells. Those new
water-treatment systems were working—reducing PFOA levels to below
levels that even the improved analytical methods could detect in the
water. The decades of drinking-water exposure were finally at an end in
the homes of the class members. But PFOA was being discovered in
drinking water in a number of other locations, including central Parkersburg, Minnesota, and New Jersey, and people in all these places were
continuing to be exposed.

In an effort to emphasize that point, I helped coordinate new water-
and blood-sampling programs for small groups of plaintiffs in communities with elevated exposures to PFOA (and other PFASs). I sent more
letters to EPA and the state agencies, documenting the individual and
cumulative PFAS blood levels in those communities and urging them to
address the contaminated-water issue and set appropriate drinking-water
standards for all the PFAS contaminants. Though it was my responsibility to advocate for the growing number of clients I had taken on around
the country, I could no longer pretend that my interest in this was strictly
as an attorney pursuing a court case. I had become that person I had
once looked at with puzzlement, if not bemusement: I was a man with a
passionate cause, a fighter against what I saw as a massive public health
threat. Somehow I had ended up trying to save the world, or at least a
part of it.

My boys took to calling me the Lorax, which I considered an improve-
ment over "Daddy eats doughnuts at work"—which was how they described
my livelihood in their early years. They regularly took turns coming
downtown to my office for a visit and for "special lunch." We would go to
Skyline for coneys or to Hathaway's, a 1950s-era diner, for cheeseburgers
and fries, then return to my office, say hello to Mr. Burke, and head to the
kitchen to scavenge for leftover Graeter's doughnuts. In those good old
days, the firm brought in doughnuts every morning. Now that the boys
were well into grade school, they understood what it meant to be an advocate for clean water, the environment, and even trees. After all, I had been
reading them *The Lorax* for years. It was one of their favorites.

The 2006 PFOA Stewardship Program was not resolving any of the
issues of PFOA blood levels or water contamination. Emissions and
production were being allowed to continue, albeit at diminishing levels,
until the 2015 deadline. Even after that, once the phaseout was complete,
the existing "legacy" contamination would linger in the environment
indefinitely. A new EPA study revealed that even after the complete
phaseout, the levels of PFOA in the environment could continue to
increase. The study showed that certain polymers in consumer
products—including nonstick pans, furniture, cosmetics, household
cleaners, clothing, and packaged-food containers—broke down through
environmental degradation into PFOA. The study found that this
process was occurring a hundred times faster than DuPont scientists
had predicted. EPA researchers found this degradation to be a "significant source of PFOA and other fluorinated compounds to the environment."

But there were things that could be done immediately to ameliorate the
ongoing human exposures to the toxin. And those measures were not that
complex: set a limit; test the water; where it exceeded the limit, filter it.

The technology existed. Granular activated carbon (GAC) filters had
been designed and funded by DuPont for our class members in West
Virginia and Ohio—and by now they had been working well for years. Yes,
they were expensive to install on a water-district scale. But their cost was
minuscule compared to the potential liability of class action lawsuits and

the public costs of the potential health problems. Why should it require litigation to make companies—and the government—do the right thing?

It took forever, but my letters eventually prompted some action—at the state level.

Minnesota set a state drinking water guideline for PFOA of no more than 0.3 ppb. It was also the first state to take steps to declare PFOA a "hazardous substance" under state law. That provoked a marathon legal battle with 3M over the state's regulation of PFAS that only escalated.

The state of New Jersey adopted a far more protective guideline of 0.04 ppb for long-term drinking-water exposures. However, DuPont rejected that guideline—which had no coercive force of law—as a basis to provide any clean drinking water in New Jersey. Just as when DuPont had refused to use the 0.05 ppb class action settlement trigger for clean water in Ohio and West Virginia under the 2006 agreement with the federal EPA, they refused to use the lower New Jersey number, this time arguing that it was inconsistent not only with the science but with the existing federal EPA consent agreement.

EPA, meanwhile, started to go backward. The agency announced that it would abandon its long-delayed effort to arrive at a final PFOA risk assessment—along with the Science Advisory Board's "likely carcinogen" recommendation that DuPont had resisted so fervently. "The assessment was becoming aged and was very preliminary," EPA said. "We're starting over again."

When EPA finally issued its first-ever "provisional" guideline for PFOA in drinking water, it was disappointing. In 2009, as EPA staff was changing with the entering Obama administration, the agency released a "preliminary health advisory" of 0.4 ppb for PFOA in drinking water, only slightly lower than the 0.5 ppb number it had incorporated into its 2006 consent agreement with DuPont. It did, however, at least revise its 2006 consent agreement to drop its 0.5 ppb trigger for providing clean water to the new figure.

That level was still higher than both the 0.3 ppb value used in Minnesota and ten times higher than the 0.04 ppb level announced in New Jersey two years earlier. And there was a critical difference: the state levels had been designed to address long-term exposure (steady consumption over years), whereas the new EPA number was defined as applying to "short-term" exposure scenarios (where the exposure lasted only hours or days). EPA would continue to avoid giving a long-term number for years, saying only that it was "still working on it."

Nevertheless, DuPont seized upon the new EPA number as alleged confirmation that the levels of PFOA in our clients' water in Parkersburg and New Jersey were perfectly "safe"—ignoring the essential distinction between long-term and short-term exposure. As a result, environmental advocates called the new EPA number a "last-minute gift [to DuPont] from the Bush administration" before the changing of the guard.

The new Obama administration declared its intention of overhauling the nation's system for regulating toxic chemicals. As part of that initiative, EPA announced that PFOA (and PFOS) would be the subject of a "Chemical Action Plan" to "outline the risks that each chemical may present and what specific steps the Agency will take to address those concerns."

The Chemical Action Plan was slated for completion by December 2009. It was never completed.

• • •

New scientific findings were coming out of Great Britain, Denmark, Canada, Norway, and even the western Arctic. But no country was studying the compound as intensely as the United States. Studies were being conducted throughout the country by top universities. PFOA was showing up all over the place—in pets, seals, polar bears, osprey eggs, and a variety of animals in the remotest, wildest places on the planet. It was also close to home: in human breast milk and newborn babies. Of three hundred babies tested by Johns Hopkins, 99 percent had PFOA in their umbilical cord blood.

And it was coming from all kinds of sources, not just drinking water. The list of implicated products was ever expanding: fast-food wrappers, microwave popcorn bags, windshield washer fluid, even some bottled water. (It was eventually filtered out.) Nonstick cookware, stain-resistant carpet, and waterproof clothing were old news. PFOA was being found in crop soil, beef cattle, and locally grown produce in areas of known water contamination.

And what did it do to people? The new studies covered a spectrum of maladies. Johns Hopkins found decreased birth weight and reduced head circumference in exposed babies. A Danish study found similar low-birth-weight results. Other studies found liver enzyme damage, increased thyroid disease, and increased ADHD in kids. These findings would all be added ammunition in my fight to get clean water and medical monitoring in my new cases, but the only findings that mattered as far as the original class action was concerned would be those of the Science Panel. So I continued to review the new data and make sure that the Science Panel was getting all of it.

By 2008, at work for almost two years, the Science Panel was still busily studying the raw data from the community in West Virginia and Ohio collected through the C8 Health Project, along with all the other available health data and studies. Our settlement agreement had set no time limit on the work of the panel, but the panel members themselves estimated that they could complete the work in five years. Access to the huge amount of data collected made it possible for the Science Panel to do some of the most extensive, comprehensive, and detailed human health studies ever done, but it would ultimately cost more than $30 million, six times our original estimate, and take even longer than expected—too long for a community anxiously awaiting news. So the epidemiologists agreed to release preliminary findings to keep the community informed. Those first results did not begin to trickle out until later in 2008.

The Science Panel members made it clear that the preliminary results did *not* constitute their official "probable link" decision. The final and actionable results were not expected for three more years. But they felt confident enough to share the early data publicly, though they would have no binding effect on the panel's ultimate conclusions.

Two days before the fiftieth anniversary of Teflon's invention, on April 4, 2008, the first results of analyses conducted using data collected through the C8 Health Project appeared in the news: "significantly elevated" levels of PFOA in the blood of nearly seventy thousand residents. The median of the entire group was 28 ppb—six times the blood levels found in the general population, according to the various blood bank studies. The highest levels were in Little Hocking water customers, who averaged 132 ppb. One Little Hocking resident tested as high as 22,412 ppb.

Other preliminary results from the community research linked PFOA exposure to higher cholesterol levels in adults and children, increased incidence of preeclampsia (pregnancy-induced hypertension) and birth defects, and delayed onset of puberty in young girls. The Science Panel also announced preliminary results finding no associations between PFOA exposure and diabetes, miscarriages, or preterm births.

"DuPont supports the work of the Science Panel," a spokesperson said in response to early results. And then, in a confounding twist of logic, added, "It is our position, based on the weight of scientific evidence, there is no human health effects related to exposure to PFOA and no risk to the general public."

. . .

While EPA delayed setting a long-term drinking-water guideline and DuPont used the new short-term number and EPA consent agreement incorporating that number to defend against our claims in Parkersburg and New Jersey, the heat was turning up on 3M in Minnesota. Minnesota's environmental protection agency, the MPCA, asked 3M to pay for the investigation and cleanup of landfills in the Twin Cities area where it had dumped PFAS. Using the familiar consent-order approach, 3M eventually agreed to pay the Minnesota agency $13 million to clean up former disposal areas suspected of contributing to the drinking-water contamination.

Meanwhile, the University of Minnesota released a new study of death rates among PFOA workers at 3M. The findings: elevated rates of stroke and prostate cancer. 3M's response will sound familiar: "Nothing in this study changes our conclusion that there are no adverse health effects from PFOA."

Weeks later, 3M's PFAS products were found in one hundred more private wells in and around the Twin Cities, expanding the number of potentially affected residents to as many as sixty-eight thousand people.

Just before the end of 2010, the Minnesota attorney general sued 3M in state court, seeking recovery for massive natural-resource damages caused by widespread PFAS contamination in soils, sediments, fish, air, and groundwater across the state. The state used many of the documents I had uncovered in our own case against 3M to make its case.

During this time, other countries were taking action to protect their citizens from PFOA. I knew that because part of my daily routine included checking and monitoring all PFAS regulatory activity online from around the world, and whenever some foreign governmental agency found the record of evidence linking PFAS to human disease persuasive enough to do something about it, I promptly sent the report to our own agencies. I was hoping to shame them into action. "See? Others understand the problem and are doing something. Why aren't you?" Of course, I also copied the Science Panel.

The big news now was that Canada set itself up to become the first country to ban the import of products containing PFOA. Norway was proposing a European ban on PFOA in consumer products.

In the United States, what was EPA doing?

· · ·

By 2006, I was juggling multiple PFOA cases in a half-dozen states, working with at least a dozen different law firms and lawyers as co-counsel, including my old West Virginia colleagues. I dove into each new case with what I thought was a reasonable expectation: that the same data, evidence, and logic that had proved successful in our original class action would apply in these cases.

Once again, I was wrong.

Over the next five years, the new cases would unfold in ways that continued to shock and discourage me. The judges overseeing these cases would reach very different interpretations of the law. In the case we brought on behalf of the residents drinking contaminated City of Parkersburg water, we filed a complaint in state court in Wood County, West Virginia, seeking the same remedies we had sought in our original class action: clean water and medical monitoring for people drinking contaminated water.

But this time DuPont had a new legal weapon to use against us: a recent federal law called the "Class Action Fairness Act." This law, which many felt had been designed to force more state class action cases into the federal court system, had been supported by industry because federal courts were typically viewed as much more "defense-friendly" than the state courts. DuPont used the new law to force our new case out of West Virginia state court and into federal court, then promptly argued that we could no longer seek remedies of clean water and medical monitoring under the rules governing certification of a class in the federal court system. Judge Goodwin, the same federal judge who had ruled on the Tennant case, agreed.

It was not because our case lacked merit. Judge Goodwin acknowledged that we had "presented compelling evidence exposure to PFOA may be harmful to human health, and the evidence justifies the concerns expressed by plaintiffs in this case." He added, "The fact that a public health risk may exist is more than enough to raise concern in the community and call government agencies to action." But according to Judge Goodwin, the medical-monitoring claims we were pursuing, based on an increased risk of disease throughout the community, were inherently individual in nature, according to federal court precedent, which differed from the state precedents under which we had prevailed in the West Virginia class action. In the judge's view, we could not show under federal case law how a medical doctor's recommendation for medical monitoring would necessarily be the same for every person in the class, each of whom purportedly had a different medical history and a different amount of water consumption.

Soon thereafter, in our New Jersey case, a federal judge followed Judge Goodwin's reasoning and refused to certify our class action for medical monitoring under the federal rules of class certification.

In Minnesota, we never even got to the issue of certifying a medical-monitoring class—the Minnesota state court refused to even acknowledge the basic right to pursue a medical-monitoring claim under Minnesota law and summarily dismissed all the medical-monitoring claims.

A similar certification glitch derailed the national Teflon cookware class action. A federal judge in Iowa denied certification of the proposed nationwide class, which included named plaintiffs from twenty-three states. Again, it was not a judgment on whether Teflon cookware was a health risk; it was based solely on the difficulty of defining who the potential class members might be. The Iowa court agreed with DuPont that because it is not always possible to tell whether a particular piece of cookware was made with a Teflon coating, it would not be possible to identify who was actually in the proposed class of purchasers of such cookware.

I was becoming increasingly frustrated and dejected. I could feel the confidence I had gained in the earlier cases draining away, one defeat at a time. But even with these setbacks, just our looming presence in the courts and our continued funneling of information and documents to the state and federal agencies accomplished some good. 3M agreed to provide funds for filtering PFOA (and PFOS) out of the impacted public water supply of Oakdale, Minnesota, after we sued the company in Minnesota. And despite DuPont's continuing insistence that the PFOA levels in New Jersey water were perfectly safe, we were able to use our lawsuit there to secure DuPont's agreement, through another class action settlement, to pay $8.3 million to cover the costs of providing water filters to our clients in New Jersey.

But DuPont steadfastly refused to provide for clean water for the City of Parkersburg.

In the meantime, clients were calling more and more often to report new illnesses and even deaths of family members. One of those reports came from the family of Earl Tennant.

Della called me, crying, with the sad news. On May 15, 2009, eight years after his case had been settled, Earl had died "suddenly" of a heart attack at sixty-seven. At least it was sudden according to his obituary. But we all knew a different truth: Earl had begun a long, slow decline when his cows started to die, and his health had never recovered. The heart attack was merely a sad endpoint, a full stop to his passionate commitment to a cause that he hadn't been able to see through to the end.

I couldn't believe that Earl was gone. I had never had the chance to say good-bye. It devastated me to think how hard he had fought to get restitution from DuPont and, once the case was over, how little time he'd had left. He would never see the end result of everything he had started. He would never know how many people he would ultimately help. Where was the justice in that?

Earl had been more than a client. Our relationship had been complicated and very hard to describe. Trying to explain what he meant to me is pretty far outside my comfort zone. But I owe it to him to try. Because Earl mattered.

He was not only the first client I brought to the firm but one of the first people who relied on me for help. I was grateful for the chance to help him. At that point in my career, I know he helped me, too. After feeling like an outsider for most of my life, I finally felt needed and valued.

Earl reminded me of my family and reconnected me to my roots. It felt good to return and feel grounded in a place that was one of the few constants of my childhood.

Earl was a man who belonged as much to the land as the land belonged to him. I admired his instincts about the natural world and his unwavering faith in those instincts, even in the face of endless doubters. The more people doubted, dismissed, and mocked him, the more stubborn he became.

I could relate to that. I was still not completely used to my new identity at the plaintiffs' bar. I still sensed that some people saw me as a defector from my tribe. I could see it in the eyes of the defense attorneys as they looked at me from the side of the table where I had once sat, making very similar assumptions about plaintiffs' lawyers myself. I even saw it in the faces of some of my fellow "plaintiffs' counsel." But the more they scowled or doubted, the more stubborn I became.

Just like Earl.

31

SHAKEN

May 2010
Crescent Springs, Kentucky

I wasn't a dark horse anymore, but I grew increasingly worried that I was becoming a dyed-in-the-wool black sheep. After the class action settlement in 2004, I had jumped into the new PFOA cases full of enthusiasm and optimism and with the full support, encouragement, and trust of the firm and my partners. Several young lawyers had devoted their fledgling careers to helping me. Thousands upon thousands of hours were being built up in those cases, and enormous sums of money were being spent on experts and litigation expenses as the cases dragged on. This was in addition to all the time and money I was still spending to help keep the Science Panel and probable-link work in the original class action on track. Every month that passed, I received a stark reminder of just how much the firm and my colleagues were investing in and counting on me in the form of billing statements with ever-increasing total hours and expenses—all flagged as unbilled and "unrecovered."

This was occurring in the context of the world's biggest economic collapse and meltdown since the Great Depression. Compared to the millions of people who suffered severe deprivation, the fiscal woes inside a corporate law firm were trivial. But that doesn't mean they were unnoticeable. I watched as longtime staff and employees left. Young attorneys who had spent much of their time on my unbilled cases left the firm. Again, nothing was ever said aloud, but in my heart I believed they had seen working for me as a deadweight hung around the neck of their careers and bolted for the door.

My fragile sense of feeling accepted in the firm, never quite secure, was eroding. As some longtime partners retired or left, new mergers with firms in other cities were announced. New partners with new business were joining the firm. Each month, dozens of lawyers who had not been at the firm when we had started the PFOA cases now looked, likely with dismay, at those billing memos with huge amounts of unbilled time and expenses next to my name.

I worried every day that my work and my practice would be seen as an unsustainable drag on the firm during a time when excessive or unnecessary expenses were no longer tolerable. It had been almost six years since the record fees from the class settlement. What was I contributing now? How was I helping my partners? I seemed to be just costing them more and more money. It was small comfort to think that I'd dragged the firm so deeply into this morass that it had no choice but to let it play out, hoping I'd eventually win and recover the losses. But it could also decide to stop throwing good money after bad. Wouldn't the firm be better off just cutting its losses at this point and asking me to move on?

If that happened, what would I do? How would I support my family? I had no "paying" clients. I had spent the last decade working for plaintiffs against the kinds of big companies a firm like mine usually defended. Suffice it to say that by this point, my phone was not ringing off the hook with corporate clients— or partners with corporate clients—wanting me to help them. I was forty-five years old with no "book of business"—how would I start over?

At night, instead of sleeping, I'd lie awake, feverishly gaming it out in my mind. We'd have to sell the house—pull the boys out of private school—cancel the lease on our minivan and turn it back in . . . I worried about these things until dawn gave me permission to get up, then kept worrying about them all day. My anxiety level was palpable. I wasn't able to participate fully in my family life because of the constant worry that the Science Panel report wouldn't be favorable, that I would never be able to get relief for my class members, that my work had become a financial drag on my firm. I realized with some melancholy that my career and this case had drained me emotionally. The things Sarah most wanted from me, I didn't have to give. That's why she sometimes scoffed at conversations about the case, which made me angry. Sarah tried to convince me that the firm wouldn't fire me over this, but I didn't believe her. I worried about it every day, especially as the economy tanked further. For now, I thought, at least I had a job and a paycheck. We had a nice home, and Sarah didn't have to work. Still, I felt like I had risked literally everything for this case and had to live with the reality that it all might blow up in my face at any moment.

It had been five years now since the Science Panel's work had begun, and despite its earlier assurances, it still had no answers. Its failure to confirm the definitive links to human disease allowed DuPont and 3M to argue that the link to human disease was still uncertain, thwarting our ability to secure relief in Minnesota, New Jersey, and Parkersburg. Those were communities we had promised we could help. Massive additional expenses were piling up. People were continuing to get sick. Some were dying. Wasn't it all my fault? After all, I had helped set up the whole Science Panel process. Not only was it taking forever, but DuPont was using it against us, claiming that any regulatory action on PFOA remained unnecessary and premature, pending completion of the Science Panel process. We were, in essence, caught in a scientific limbo by the very process we had worked so hard to bring about.

· · ·

I sank deeper and deeper into despair. I'd spent almost a decade tilting at the PFOA windmill, and where had it gotten me or my clients?

Were others seeing the handwriting on the wall, too, and worried about being connected with me and my work for fear it would end their future at the firm? It seemed as though the number of people walking past my door or stopping by to chat was rapidly declining. Rarely did anyone call or invite me to lunch anymore. Was I just growing paranoid, or were people avoiding me now?

As a kid living near military bases where either you or your friends moved every other year, I had learned to protect myself from being hurt by being the one to withdraw first and to stay distant so as not to be surprised or upset when a relationship ended. I often stayed in my office with the door shut. Maybe I was responsible for my own isolation. Maybe. In any case, I knew I had to prove to the firm that its trust in me had not been misplaced and I deserved my place there.

I had the ability to work harder and persist longer than anyone I knew. Even when things looked hopeless—and they'd been looking that way for quite a while—it just wasn't in my nature to give up. The only thing I could do about it was work even harder. My own work was the only thing I could control and the one place where I had ever felt confident.

I knew I had uncovered a terrible truth. I knew it was dangerously real. It had been successfully buried for all these years, with complex systems designed to obscure it. I knew that billions of dollars were at stake if the truth should be fully realized and that there were those who would fight viciously to keep it hidden.

Human life was at stake. That was what I had to defend. If the full scope of the problem wasn't exposed, no one—at the corporate or regulatory levels—would be forced to change their behavior. People would continue to be unknowingly exposed. People like Earl and Joe Kiger and even Sarah's mom would continue to get sick and wonder why. People would continue to die too soon.

This might all sound like conspiracy talk. But I now believed the efforts to conceal the truth were real. I knew what I was up against. Both DuPont and 3M had long-standing corporate reputations that seemed incorruptible. And they had great incentives—and resources—to defend their reputations. They could afford the best lawyers, consultants, and spokespersons, and those professionals were very good at their jobs. Their statements sounded so reassuring, so convincing, and so logical that it was no wonder that people believed them.

Some days it all seemed insurmountable. But every time I grew despondent, I'd think of Earl. He had pushed me to keep fighting until DuPont was finally held responsible. He hadn't been fighting for money, fame, or revenge; he had been fighting for accountability. Earl had never second-guessed himself. And, in the end, he was right.

. . .

Back in September 2008, we had lost Grammer. Her death hadn't been entirely unexpected—she had been ninety-one, suffering from lupus, and confined to a wheelchair at my parents' house—but her final decline had been sudden and swift. I had spent weeks driving back and forth to Dayton to visit her in hospice. And then she was gone.

Soon after, Jim and Della Tennant's oldest daughter, Martha, was diagnosed with breast cancer. After a double mastectomy, she went through chemotherapy. I didn't know Martha personally, but Jim and Della had told me a lot about her. She was a mother of two young kids. Earl's wife, Sandy, also was battling cancer now. It wouldn't be much longer before she succumbed to the disease and passed away. Like Earl, far too young.

Jim and Della still called me all the time and had become true friends. Every Christmas, I looked forward to the box of homemade fudge Della would send us, along with the wonderful notes, emails, and cards from both of them throughout the year. Just hearing their voices on the other end of the line and their unwavering support and encouragement meant the world to me. I couldn't imagine what they must be going through.

At home, Sarah was dealing with the resurgence of her mother's cancer. My parents' own health was declining. I was dealing with a series of heart attacks and other maladies suffered by both my mom and my dad. Even my sister had a heart attack at a ridiculously young age that shocked everyone.

During all this, the Science Panel continued to bog down in seemingly endless delays. The data analysis had proven even more difficult than it had expected, and it had announced several pushbacks of its target dates. By design we had no say in its timeline and both sides were strictly prohibited from getting involved in the panel's methodology, short of providing whatever the three panelists asked for. They assured us they had all the consultants they felt they needed. This was complex, rigorous science, and they knew that whatever results they announced would be picked apart by a phalanx of high-paid attorneys. It had to be bulletproof. That simply took time, apparently a lot more of it than we had expected or were comfortable with. Ed and Harry were fielding complaints from increasingly irritated class members. And of course DuPont and 3M continued to wield the delay like a club—"no evidence of human health effects"—and with success.

The frustrating thing was that the Science Panel's long delay was all due to the massive data gap—the lack of any existing complete epidemiological studies on PFOA among exposed community members. That gap couldn't be a fluke. I had noticed a pattern within the epidemiological research conducted by DuPont and 3M. They had initiated human health studies on PFOA. But when the preliminary results had indicated possible adverse effects, or at least the need for further investigation, many of the studies had stopped—like DuPont's worker pregnancy study, which had been "put on hold until further notice" in 1981. Or the studies weren't published, or the conclusions in the published versions of the study often did not correspond to the raw data. Some studies were never conducted at all. I thought about the twenty-year gap between the 1978 monkey study—in which all the high-dose monkeys had died—and the subsequent primate study in 1999. Where were the follow-up monkey studies in between?

All of these thoughts and questions swirled through my mind nearly every waking minute of every day. I was sometimes irritable or distracted, but I tend to internalize pressure. Just about the only real relief I had was watching episodes of *Family Guy* with our cats, Kong and Possum (that and stress eating, which had packed almost twenty pounds onto my frame). I don't know what it was about *Family Guy*—the dark humor, maybe, or the political satire—but it made me laugh so hard that I sometimes scared the cats off my lap. (They always came back.)

One Sunday, while Sarah took the boys to church, the cats and I had settled onto the couch for some cartoon therapy, and I was able to disengage my brain from PFOA for at least a while. That was when I first started to feel a little strange. My vision seemed to be blurring, with odd distortions and bursts of light in the corners. I shook it off and went upstairs to get dressed for the day. But in the shower, my vision got worse and I started feeling light-headed and dizzy. I figured it was just the steam. Then, while walking down the stairs, my feet went numb.

I groped my way to a chair in the kitchen as the right side of my body began to tremble. The tremors took over my right hand and right leg. I tried to put my sock on—and couldn't.

Sarah and the boys found me sitting there, shaking. My speech was slurred, and the tremors had escalated into a palsy on my right side. As I tried to respond to Sarah's questions, I realized that I now couldn't speak at all. I started to panic. Sarah helped me move to the living room couch, where I began to black out. She sent the boys downstairs so they didn't have to see me like that. Then she ran down the street to get help from Mark Boyd, a doctor who lived on the corner of our cul-de-sac.

The next thing I knew, Dr. Boyd was standing in front of me in our living room, looking worried. He made me lie down on the floor and prop my feet on the couch. I didn't pass out, but the palsy in my right leg and arm became uncontrollable. He shined a light in my eyes, and I could guess what he was thinking: *a stroke.* He didn't say it, and I'd lost the ability to speak, but we were thinking the same thing. He called 911.

As the paramedics wheeled me out our front door to the ambulance, Sarah sent the boys to stay with Mark's wife so she could come with me to the hospital. I saw my son Charlie start to cry as he joined the other two running down the street. I was horrified that they had seen me like this. I was humiliated when I saw the neighbors up and down the street watching me shake my way into the ambulance with its lights flashing in our driveway. Then, in the emergency room, the palsy turned into violent shaking convulsions all up and down the right side of my body. I could not stop my arm and leg from shaking and banging repeatedly up and down on the emergency room table. This went on nonstop for three hours. Sarah held herself together with remarkable stoicism all that time. As the doctors strapped me down on the table, I saw the fear written on her face. And just like that, the shaking stopped. As though someone had flipped a switch. My vision was still foggy, and my mind was fuzzy. I could hear people talking, but I couldn't respond.

I was frankly terrified. Then it passed, and I just felt angry at my own physical weakness.

It wasn't a stroke. And none of the doctors had any explanation for what it was. The EKG said it wasn't a heart attack. They did a brain scan, an MRI, all sorts of tests, but still no diagnosis. I stayed in the hospital for several days, waiting for answers that never came. It drove me crazy, not knowing what the hell it was or what had brought it on.

A seizure? Some weird neurological disorder? No one could tell me. The uncertainty was as disturbing as the symptoms.

It took several days, but eventually the symptoms faded and I started to feel something like normal. I went back to work, and things seemed fine. At least for me. Sarah was furious that I wasn't taking it all much more seriously. She was convinced that I was under too much stress and things were just going to get worse if I didn't slow down and let others take the reins on the DuPont business. She grew increasingly angry thinking that the firm and my co-counsel were making me sick, putting too much pressure and stress on me to do everything myself and to feel responsible for everything that was—and wasn't—happening. More than once, she threatened to call my colleagues and tell them exactly what she thought. She and the doctors urged me to schedule more appointments with more neurologists. My parents and sister echoed her concerns. But I decided I would simply chalk it all up to some weird "episode" and move on and forget it. I didn't have time for all the fuss.

I was still writing letters to EPA and working on our new PFOA cases, but my partner, Kim Burke, had asked me for help on one of his cases. It felt good to be back, however briefly, in my original habitat.

Then it happened again. If the first time was scary, the second episode was horrific. This time, I was at the firm.

I had just returned from taking a deposition in Kim's case. I walked into my office, sat down at my desk, and watched as my computer screen grew blurry and wavy, as if it had turned into liquid. My head grew foggy, and I started getting dizzy. Then my right arm and leg started shaking.

I tried to get up, but I couldn't. I was panicking as I worried that the shaking would turn into unstoppable convulsions again—and then what would I do? I had to get out of there—quick. I hit the button on my phone to call Deborah, my assistant, to tell her I was taking off early. When she answered, I realized that my speech was slurring and I was having difficulty getting any words out. It was too late.

When she heard my garbled speech, she raced down the hall and opened the door. As soon as she saw me sitting there with my arm and leg shaking hard, unable to respond to her questions, I could tell she was thinking I was having a stroke. She didn't know about my previous episode, because I hadn't told anyone at work. She dialed 911.

The medics arrived, strapped me down on a stretcher, and wheeled me through the firm. I was convulsing uncontrollably, but I still had the presence of mind to notice everyone staring—in the hallways, on the elevator, on the street. It was getting close to 5:00 p.m., and people were everywhere. It was absolutely humiliating. For someone who hated being the center of attention, to be the object of pity was the worst kind of attention I could imagine.

Again the convulsions lasted three hours. The doctors called in another neurologist, who did an MRI and stuck electrodes all over my body. The shaking started again. At least, I thought grimly, they'd be able to figure out what was happening to me and fix it. "Unusual brain activity" was all they could tell me. It wasn't a seizure, a stroke, or some form of epilepsy. They used the analogy of a computer with a wire shorting out. But short of doing an autopsy (I respectfully declined), they said, "We may never know."

. . .

At this point in my life, insult and injury were conjoined twins.

I tried to hide them. Once, when I felt another episode starting at work, I tried to hurry out of the office before anyone could see me. I didn't get far before my right leg started buckling beneath me. It felt as though all the muscles in my leg simultaneously went limp. Because it would happen with no warning, I had to catch myself to stop from falling over. I had to very carefully trudge back to the firm and glumly ask Kim Burke to drive me home.

Being embarrassed at work was bad. But knowing I was adding to the stress and worry of my partners, and especially Sarah and my family, was unbearable. Sarah was busy enough with three boys at home and a mother undergoing chemotherapy. I hated to think I was creating any additional issues for anyone. I would sit on my hands in meetings if I felt the slightest tremble. If my speech started to slur or my leg gave out and I stumbled, I worried that people would think I was drunk. Especially at first, the affliction made me withdraw even further into extreme isolation. After a quarter century in the downtown Cincinnati office, I chose to move across the river to the firm's Northern Kentucky office (nearer where we lived), where there were only three other attorneys and one assistant. I had now physically disconnected myself from the office where I had spent my entire career.

I was feeling increasingly disconnected from my co-counsels as well. It seemed that no one else fully understood how complex and multilayered the whole PFOA issue had become. Certainly, they could commiserate about battling DuPont or 3M and the excruciating wait for the Science Panel to finish. But they worked in other states and had lots of other cases, and we rarely met in person.

With multiple cases—in Minnesota, Parkersburg, and New Jersey, all still active at that point—and my ceaseless monitoring of new PFAS research, writing letters to the federal and state agencies, following regulatory actions around the world, and keeping track of whatever DuPont was feeding to the Science Panel, I felt like a one-armed man directing a three-ring circus.

I determined our battle plans and tactics, organized and coordinated our forces. In this instance, many of my foot soldiers were plaintiffs' lawyers, who are famously self-confident and prone to think they can do it all. They didn't know what they didn't know. On the other hand, I knew enough to be painfully aware of what I didn't know and was constantly focused on building a complete team, to bring in the people who would fill the gaps in my knowledge: attorneys experienced in class actions and others intimately familiar with the West Virginia, Minnesota, or New Jersey courts and personnel and well connected to the local community; an analytical chemist to decipher water sampling; a world-renowned toxicologist; an epidemiologist and a physician to teach me about epidemiology; a public-health expert. The list—and necessary team members— just kept growing. Though each of the team members focused on discrete tasks and functions, there had to be one person keeping all the balls in the air, to grasp the big picture and monitor the seemingly never-ending attempts to influence the science, the regulators, the media, the politicians. That was, inescapably, me. As the complications metastasized, keeping everyone up to speed became increasingly difficult. The stress of monitoring all of these things simultaneously was magnified by the certainty that if I slacked off in any one area, that would be exactly where DuPont or 3M would attack. I was always outnumbered—me and a dozen part-time attorneys against the in-house counsels of two Fortune 500 companies, plus the half-dozen huge national and local law firms they'd hired to assist them.

Being so intensely focused on the details made it difficult sometimes to back away and find the perspective I needed to answer the big question behind everything: How had this happened?

How does a chemical pollute the whole globe? How can something that kills animals and lurks in human blood for decades remain undetected for so long? How can a company contaminate the water of entire communities and get away with it? How could regulators—whose sole purpose was to prevent such things—stand by and let it happen?

I felt an enormous responsibility to bring the truth to light. This wasn't just some case about environmental permits and regulations; it was about people's right to turn on their tap and trust that the water that fills their glass is not tainted with a chemical that could hurt them.

The doctor who examined me after one of my episodes asked, "Are you under a lot of stress?"

I told him, "Nothing more than usual." I think Sarah could have strangled me.

. . .

As 2011 rolled around, we were still waiting on the Science Panel. The community was grumbling, and the court was giving the panel heat, which was echoed in the press. Judge Hill had died in December 2011, at the age of eighty-one. A new judge, John D. Beane, had been assigned the responsibility of overseeing the ongoing settlement activities. Though we couldn't push the panel directly, we could tee up a hearing with the judge, who felt as antsy as the rest of us.

"I'm just so frustrated," the judge told the *Parkersburg News and Sentinel*, referring to the panel. "You're getting paid well to do this, and I just don't see anything being done."

He even threatened to require both parties to appoint a new Science Panel unless things picked up. He never went that far, though he did order the panel's members to appear in court and explain what was taking so long.

"The public wants answers," he said.

Kyle Steenland, one of the Science Panel members, was quoted in the *Parkersburg News and Sentinel*, explaining the delay: "We need to be able to estimate how much PFOA people have been exposed to in the past. If someone had cancer in 1995, we need to know what their exposure before 1995 was."

This was Olympic-level epidemiology. Using DuPont's historical records of PFOA emissions into the river and air, the panel members were modeling where the PFOA had gone and calculating how much was likely to be in each water system at different times. Factoring in when and how long an individual had consumed water at those levels, they were able to estimate each person's individual exposure and even calculate estimated historic PFOA blood levels for him or her.

"This is a huge task, yet it is key to making a solid scientific decision about whether C8 is linked to disease," Steenland said. "We have always said our work would take time, but it is worth it to get it right. We know the community is eager to get results, and we are working to get them as soon as we can."

The Science Panel did continue to release intermittent preliminary findings, including increased rates of death from kidney cancer and kidney disease among the DuPont Washington Works employees exposed to PFOA. In nonworker residents, they found PFOA levels associated with increased liver enzymes and an association with preeclampsia. But they continued to emphasize that these were simply preliminary "associations," not necessarily "probable links."

The class members' impatience threatened to boil over into anger. Joe Kiger was vocal about his frustration with the long wait.

"I'm seeing a lot of sashaying around the block," he told the paper. "They keep coming up with association after association, but they haven't said 'probable link' yet."

The Science Panel's preliminary findings of certain associations were mildly encouraging. But it had announced the absence of many associations, and those were even more numerous. The longer it went without naming a single probable link, the more I worried that it wasn't going to.

Apparently the Science Panel was feeling the pressure, too.

"We take this very seriously," said Science Panel member David Savitz. "We recognize the stakes in this community, the financial stakes, the emotional stakes, and the public health consequences."

DuPont, not surprisingly, seemed content to wait, and the longer the better. "We have not pushed for a timetable on this," said Larry Janssen. "We have full confidence that if they find something, they would tell."

And the Science Panel members had full confidence that if they did find something, they would be subject to an inquisition that no amount of excessive care on their part would forestall. "We've come to realize that basically every decision we make will be challenged," Savitz said. "The process is being handled properly and was thought through carefully, but at the end of it, there will be varying degrees of controversy and disagreement that I'm afraid will remain."

This agony of uncertainty finally came to an end in September 2011 with an email from the panel setting up a conference call. With our team and DuPont's on the line, it announced that it was expecting to make a probable-link announcement very soon. I was nervous and excited. Under the terms of our original settlement, the finding of even one link would trigger the formation of a new Medical Panel charged with overseeing any future medical monitoring for the exposed class members. It would also lock down DuPont's obligation under the settlement to continue paying for all the filtration systems cleaning the community's water— forever.

I was eager to move forward and already preparing for the next steps. Within days of the call, I had a series of neurological episodes—one requiring a hospital visit in October, then three in a row in November. Same experience. Same symptoms. Again, no one could explain them. The doctors simply switched me to a new combination of meds. Dealing with my mystery condition was a long, disquieting process. Early attempts at medication left me thinking through a dense mist in my brain and ate away at my ability to sustain focus. I told the doctors that wasn't going to cut it. Clear thinking and sustained focus were all that I could bring to the work I did, which was hard enough when I wasn't in a medically induced fog. I went through a variety of different medications prescribed by an ever-changing lineup of doctors. Eventually we found a combination of meds that seemed to minimize the "episodes." But I still had an embarrassing and unpredictable tic. My head would suddenly jerk to the side, one eye would wink closed, or one arm or leg would spasm. My leg might occasionally go limp as I walked. It made me extremely self-conscious.

The episodes continue to this day, though they have grown less severe. Few days pass when my leg doesn't suddenly go out or my neck chooses to spasm violently or my arm jerks and eyelid twitches. But I can still do the work, so I live with the rest.

32

ROAD TO A RECKONING

Winter 2011
Vienna, West Virginia

On December 4, 2011, the Science Panel released its first probable-link report. It had found *no* association with preterm birth, low birth weight, miscarriage, or birth defects. I was floored on the birth defects result. What would we tell Sue and Bucky Bailey?

But it did confirm a link: with preeclampsia—pregnancy-induced hypertension with serious implications for the mother, up to and including death.

I have never been so relieved in my life. After all those years and all the effort—independent scientific verification that PFOA was linked with serious human health effects. No more arguing with DuPont over whether there was "no evidence of human health effects" at the "trace levels" of exposure to PFOA in the community. The issue had been resolved once and for all—and with independent science.

Yet that relief was short-lived, as DuPont immediately began questioning the significance of the Science Panel's finding through a corporate spokesperson, despite the fact that they had helped select and pledged to support the panel. Because I knew my opponent so well, I had nailed down the terms of the settlement to compel both sides to accept the findings of the panel without any means of challenging them. Adding insult to injury, DuPont had been using the panel and the slow pace of its painstaking work for years as a shield against new lawsuits and an excuse to do nothing about their ongoing pollution, and now that the panel's results weren't to their liking, they were discounting the significance of the results.

The *Parkersburg News and Sentinel*, echoing DuPont's message, dismissed the announcement as "frankly, underwhelming." Here's what the editorial said: "So far, it seems the potential dangers of PFOA exposure have been greatly exaggerated, with fears fanned by some members of the legal community whose interest, we suspect, goes beyond worries about the health of area residents."

The term "money-grubbing lawyers" was absent but implied.

DuPont could scoff all they wanted, but over the next several months, the Science Panel trickled out the rest of its findings. At the end of July 2012, it announced its final set of results, more than seven years in the making. Since the first link had been announced in 2011, it had now confirmed probable links between PFOA exposure and six diseases. It seemed that the potential dangers of PFOA had not been "greatly exaggerated" after all. The list went like this:

> Kidney cancer
> Testicular cancer
> Ulcerative colitis
> Thyroid disease
> High cholesterol
> Preeclampsia

There it was; our burden of proof had been met. Three independent, world-class epidemiologists had studied one of the most comprehensive, complete, and accurate data sets in the history of epidemiology. They had been given unlimited time and funding to take every measure they felt they needed to produce unassailable results. They had reviewed and evaluated all the other existing studies—published and unpublished—and all the data. After putting it all together and weighing all the evidence, they had found probable links with six diseases. Two of them were cancers.

Only now that the anxiety over the verdict of the Science Panel had been lifted did I fully realize how much it had worn me down over the past seven years. During that time, I had watched my sons progress from peewee soccer to high school sports, spending less time with them than I should have. All those years, DuPont and 3M had been calling us liars, accusing us of using scare tactics to alarm and mislead people. Finally, we had proven them categorically wrong.

Now that I could stop obsessing about the Science Panel, we could begin to focus on phase two of the settlement. We would have to begin the tedious and slow process of figuring out the best way to formally notify seventy thousand class members of the Science Panel's results and explain their rights. The finding of probable links meant that the water-filtration system would remain operational forever. It also meant that medical monitoring would now be made available to our class members. As specified in our settlement agreement, we would now select the Medical Panel that would determine the details of our medical-monitoring program. We picked the Medical Panel through the same process we had used for the Science Panel. Instead of epidemiologists, though, we were looking for three unbiased, world-class, independent physicians. And this time, Larry Winter wasn't available, so Larry Janssen had to travel with me.

My concerns about our personality conflict turned out to be overblown. I'm sure I had been right in thinking he'd regarded me with disdain when we were still adversaries, but now the case was essentially over as far as he was concerned. We were implementing the settlement he had helped negotiate. There was nothing left to fight about.

Even so, given our history, things were tense early on, until one evening on the road we had dinner together for the first time in an informal setting. The case was never mentioned. Instead we got to talking about how our firms were struggling to adapt to a changing economy and discovered we had very similar experiences. For the first time, he was able to see me as one of his tribe, a corporate defense guy.

We didn't become best friends after that, but our relationship remained unfailingly cordial throughout the four-month process.

. . .

By April 2012, we had finalized our Medical Panel choices: Dr. Dean Baker from the University of California, Irvine, Center for Occupational and Environmental Health; Dr. Melissa McDiarmid from the University of Maryland School of Medicine; and Dr. Harold C. Sox from Dartmouth Medical School.

Under the settlement, the Medical Panel was required to accept as proven and established that PFOA was capable of causing the six diseases the Science Panel had confirmed among the class members. For medical tests and procedures to be the financial responsibility of DuPont, the panel would have to decide if they met certain conditions, including:

1. Tests for the diseases existed.
2. The tests required were different from what doctors would normally recommend for an unexposed patient.

In addition, the panel would determine which class members faced a sufficiently increased risk of the disease to make individual class members entitled to these tests. DuPont argued that only those in the highest PFOA-exposure groups within the class had sufficiently high enough exposures to warrant monitoring. I argued that the Science Panel had found the probable links to exist among the entire class, and thus anyone who had enough PFOA exposure to be a class member should be able to get monitoring. That was a critical point. If DuPont won that argument, they could effectively restrict the availability of the medical testing to only some small fraction of the class members who had the highest exposure levels. We ultimately won that battle.

In May 2013, the Medical Panel released its initial report. Other than certain obvious distinctions based on age or sex (for example, there would be no women monitored for testicular cancer or men monitored for preeclampsia), the panel recommended that, in essence, everyone in the class ought to be monitored for all six diseases. Under the terms of the 2004 settlement, DuPont was now required to pay for all such monitoring, up to a maximum of $235 million.

Though the settlement had spelled out the terms for medical monitoring, what it did not clearly spell out was exactly *how* to go about monitoring seventy thousand people for six diseases. Nothing on that scale had ever been done before. We planned to have Paul Brooks and Art Maher of Brookmar again figure that out. They had proven their organizational ingenuity and administrative capacity with the C8 Health Project, which had not only met ambitious goals but exceeded them. They could build upon the IT systems, infrastructure, and team already in place and had already started sketching a plan for one-stop shops where people could come in, get tested, and walk out. Their goal, as it had been in the blood testing, was to make it easy.

It seemed that DuPont had other plans. They were now very well aware of just how successful Brookmar had been in getting almost the entire seventy-thousand-person class to show up and participate in blood testing. That testing had provided the data needed to confirm the probable links. Now there was up to $235 million at stake to pay for the medical monitoring recommended by the Medical Panel. But under our settlement, we had agreed that DuPont would have to reimburse only the cost of the testing that the class members actually used. In other words, the Medical Panel could recommend the most comprehensive and expensive testing in human history for everyone, which could easily reach the $235 million cap if all seventy thousand people participated. But if no one in the class actually showed up and asked for the tests, DuPont would not have to spend one penny. Having an outfit like Brookmar, skilled and focused on helping people easily access available testing as quickly and efficiently as possible, involved in such a situation certainly wouldn't be in DuPont's best interest financially. So DuPont proposed using someone else: Mike Rozen of the Feinberg Rozen law firm in New York City.

Rozen had no medical expertise, but he was one of the foremost experts in distributing compensation to victims of mass tragedy. He had served as the deputy special master of the US government's September

11th Victim Compensation Fund, created in 2001; the deputy administrator of BP's Gulf Coast Compensation Fund after the *Deepwater Horizon* blowout; and adviser to Pennsylvania State University in the resolution of claims arising out of the Jerry Sandusky sex scandal. DuPont argued that Rozen and his firm's experience in handling the implementation of massive, complex settlements would be exactly what we needed.

As careful as I had been in the details of our settlement agreement, I realized with rising nausea that we hadn't nailed down exactly how the unprecedented new medical-monitoring program would be carried out. So once again we needed to negotiate with DuPont. But this time, their lawyers had the leverage: any long delay would deprive the seventy thousand class members of their hard-won compensation. We had people waiting for cancer testing. They didn't need it in the distant future. They needed it now.

Motivated to reach an agreement, we managed a compromise: we would agree to Rozen as long as he agreed to collaborate with Brookmar on the rollout of the program. DuPont would pay the costs of implementing the program, including a fee for Rozen. With that deal inked and filed with the court, we were ready to launch the massive testing program. But when it came time for Brookmar to move forward with its plans, Rozen announced that he didn't think Brookmar would be the right choice.

I filed a motion with the West Virginia court asking that it enforce our deal and order Rozen to let Brookmar move forward. Judge Beane set the matter for a hearing, and DuPont played a familiar card—with an outrageous twist: they promptly filed a motion to disqualify the judge. I couldn't believe it. Here we went again. This time, DuPont argued that Judge Beane could not rule on the issue because his father, who had been a local physician, had had an office in the same building as Paul Brooks decades earlier. Really? That was absurd. Judge Beane promptly denied the motion, but DuPont appealed it to the Supreme Court of Appeals of West Virginia. It was now clear to me that DuPont was going to fight to keep Brookmar as far away from the new medical monitoring program as they could.

The legal wrangling had another unexpected consequence: Dr. Brooks and his colleagues were furious that, in their view, they were being called a bunch of unsophisticated local yokels who couldn't possibly handle the advanced new program that Rozen was putting together. He had had enough. He told us that even if the court said Rozen had to use Brookmar to implement the program, they no longer wanted to work with him: either Rozen went, or he would go.

Now we were in a real mess. Rozen wasn't offering to step aside and abandon the $250,000-per-month flat fee he had negotiated for himself with DuPont. I suspected that even $250,000 a month was viewed by DuPont as a steal if it would lead to fewer people ever submitting a claim against the $235 million. It was clear that we were at an unresolvable impasse. Meanwhile, we had a medical-monitoring program ready to go and class members were waiting. I felt we had no choice; we would have to move forward with the medical-monitoring program without Brookmar. When the new notices finally went out, class members started signing up, but slowly. Eventually, more than eight thousand class members signed up for the program, which included a free PFOA blood test and doctor visit, along with a myriad of diagnostic tests for the six linked diseases. That made it one of the biggest medical-monitoring programs ever set up. Yet it was only a fraction of the number who had given blood in the C8 Health Project, and I couldn't help but wonder how many more would have taken advantage of the free testing if Brookmar had been allowed to guide the process.

· · ·

Now that both the Science and Medical Panels had issued their reports, the water-filtration systems were in place, and free medical monitoring was being offered to our class, all of the objectives of our class action had been accomplished. We could start thinking about a subset of our class members, the unfortunate ones who had contracted one of the linked diseases from drinking PFOA-laced water. Unlike in the first phase of the case, where we had represented everyone in the class as a group, these personal-injury and death claims were viewed as individual claims to be pursued through separate, individually filed lawsuits by each class member using whichever attorneys he or she chose.

The reality, so long in coming, was stunning to think about. Eventually, roughly 3,500 people within the class of approximately 70,000 (including DuPont's own workers at the Washington Works plant) chose to come forward and identify themselves as people who not only had been diagnosed with one of the six linked diseases but were now willing to file a formal lawsuit against DuPont to recover damages caused by that disease. I was sure that there were many more people in the community who suffered from those same diseases but simply did not want to risk taking on DuPont. The case was no longer solely about law and science; it was about the human misery caused by five decades of DuPont's actions and inactions. As Ed, Harry, and Larry canvassed our new clients in the individual lawsuits, they gathered heartbreaking details of their personal stories—stories that would finally get their day in court.

Stories like that of Ken Wamsley, a Washington Works employee who had worked with Teflon-related products for more than two decades.

Ken was a dyed-in-the-nylon DuPonter who had never doubted he'd be a "lifer." Like everyone else in the company, he had paid his dues and changed jobs many times. He had started out shoveling snow in 1962, when he was just nineteen. He had spent the next eleven years packing and shipping boxes of Zytel—a high-performance, heat-resistant nylon resin—to customers who molded it into car engine parts and cable insulation, among many other things.

In 1973, Ken was promoted to the casting floor in nylon. Wearing a face shield and long gloves, he handled the white-hot molten polymer that came out of industrial ovens in a viscous state. He had not loved that job. Between the heat and the industrial machines, there were so many ways to get hurt. It was dirty and physically demanding. He knew he could do better.

He enrolled in night classes at Marietta College, across the river in Ohio. DuPont paid his tuition, and he was grateful. He discovered he had an affinity for math. That opened the door to a better job. He had set his sights on laboratory work.

Ken aced the tests DuPont gave and got a job testing nylon and Zytel products.

In 1976, he made his final job change and transferred to the chemical analysis laboratory at Washington Works. There he tested Teflon products of all kinds, using high-tech machines to manipulate molecules and analyze chemical compounds. As he had come up in the company, he had heard some talk in the breakroom about working with Teflon—*you don't want to go there if you don't want to die early*. But he'd figured it was just workplace grousing on the part of a few malcontents. He had always been impressed by DuPont's emphasis on workplace safety, and anyway the new work was too good an opportunity to turn down. He loved the precision: he could identify the presence of something at infinitesimal levels—parts per million, parts per billion. Every morning, he arrived a few minutes before seven to begin the first experiment of his day: the aqueous solvent extraction of more than a thousand compounds from *C. arabica*. In other words, making coffee.

But first he had to wait for the water to run clear. Pumped from the groundwater beneath the plant, the stream emerged from the faucet with a brownish tint. He'd run the tap until the water looked like something he'd want to drink. He and his colleagues in the lab grumbled about it and complained to superiors again and again. They were told it was nothing to worry about. So they didn't.

By 1976, Ken had been drinking the Washington Works water for better than a dozen years. One day, at work in the Teflon lab, he started feeling stabbing pains in his abdomen. The cramps sent him dashing for the bathroom. At first it felt like a bad stomach bug or maybe a case of food poisoning. But he'd suffered from both before, and those pains had gone away. These persisted. Soon the cramps were doubling him over. Some days were better than others, but the bad days were getting worse. He felt comfortable enough to stay at work in the lab, where there was always a toilet nearby. But he grew anxious about going out and about. Whenever he was somewhere other than home or work, he was constantly scanning the scene to locate the nearest restroom. He had some very, very close calls.

One night the pain grew so intense that he wasn't sure he would make it through the night. His wife called the doctor and woke him up.

"Get him to the hospital," the doctor told her. "Quick."

In the emergency room, Ken was poked and prodded in an increasingly invasive series of tests. But the source of the problem was not apparent.

"We're going to open you up and do exploratory," a doctor said.

The surgery revealed ulcerative colitis: his small intestines were covered with adhesions, little bands of scar tissue that were destroying his digestion, like sand in a gas tank. The doctor excised the adhesions and sewed him back up. Ken woke up the next day to find an incision that started eight inches above his belly button and ended below his belt. His doctor's words would echo in his ears on the nights he could not sleep: "Ken, I got your problem fixed. But I'm worried about your future."

Recovery from a cut like that was hell. He lay in bed, snagged in a web of tubes draining his wounds. When he could finally leave the bed, he sat in some kind of swing chair they'd rigged for him. He kept trying to will himself to stand, but his body couldn't do it. Too weak. It took months for him to start to regain some strength. He was hemorrhaging money, just sitting there unable to work. He forced himself to walk up and down the street outside his house. He was anxious to get back to work, worried about losing his job if he took too long to recover. After about three months, he returned to the Teflon lab.

Bent over the bench, manipulating molecules in beakers and cylinders, he eventually felt as good as new. Life was back to normal. He started feeling himself again, and he was becoming one of the most experienced lab technicians in the division, the one younger colleagues came to for help.

Ken loved his job in the Teflon lab. One of the things he tested there was what the DuPonters called C8—PFOA.

Like Karen and Sue, Ken thought of the substance as being "just like soap." It made the Teflon products "more slippery." He measured the amount of PFOA various products contained. Someone would bring him a sample in an eight-ounce bottle. He'd pour it into a graduated cylinder, run it through a cheesecloth.

Whenever tests called for baking the Teflon in the white-hot furnaces,

Ken stood back from the heat and the fumes. His colleagues warned him not to breathe Teflon fumes. They told him it was because you "can get pneumonia just like that and the next day you'll be sick." Workers wouldn't be advised to wear masks until the 1980s. But Ken was careful not to breathe anything if he could help it. That was just common sense.

Safety rules and procedures had been drilled into him since his first day. Safety glasses, safety shoes, safety goggles, different types of gloves—each job had specific requirements. If the plant safety superintendent caught you doing your job without proper safety gear, you'd get an earful. You might even be sent home for the day. But that never happened to Ken. He was that guy who followed every rule in the book.

The lab manuals were extremely specific about safety procedures for handling chemicals. Ken occasionally had to work with some known carcinogens, which he handled carefully under an exhaust hood. PFOA had no such warnings, so he worked with it "right on the bench," wearing gloves to protect his hands. Once in a while something would splash on his arm, but he didn't fret. Just wiped it off. He didn't see anything in the safety books that instructed him to do otherwise.

Like every other employee, Ken saw the plant physicians for routine checkups, and from time to time they would draw his blood and test it for the presence of chemicals. He never worried about the results. He felt sure his supervisors cared about him, and he trusted that the plant's exhaustive safety policies and procedures offered more than enough protection. If anything, he thought they seemed excessive.

Of course there was the brownish water that streamed from those plant taps. That seemed an annoyance more than a serious hazard. He and his colleagues harangued their supervisors for bottled water, but they didn't get it. Brown water was the status quo, and Ken didn't lose any sleep over it.

Mainly he appreciated the fact that his work was steady and dependable. Except when the plant decided to extend the shifts from eight hours to twelve; that made the days seem endless. And it took even more coffee—sometimes six or eight cups, all made with the funny water—to get through them.

In 1980, Ken was thirty-seven years old, a husband and father of three youngsters—two girls and a boy. He'd bought a house in Parkersburg and took his family to the company picnics, where he was still young and spry enough to be a ringer in company softball games. He was fast on the base paths and had a rifle arm. Those were the happiest days of his life.

That same year, DuPont's Polymer Products division recorded the best safety performance of any department in the company, the greatest number of "exposure hours" without a lost or restricted-workday case. And within it, Washington Works claimed to have the best safety record of all 142 DuPont plants around the globe. "Washington Works is the No. 1 plant in the No. 1 safety department of the No. 1 safety company in the world," Ken's friend and colleague David Morehead would write many years later, remembering 1980.

Just months later, in the spring of 1981, Ken came into work and noticed that something was off. Where were the women? Every female who worked in the Teflon division was missing. That was odd.

He heard that all of the women had been sent home.

But why?

Someone mentioned something about a chemical that was suspected of causing birth defects. The chemical was PFOA. Ken tried not to worry; they said it affected only the women. But it did give him enough of a tilt to ask his supervisor about it. His supervisor looked him square in the eye and put his fears to rest. "Don't worry, Ken. It won't hurt men."

For twenty years, even though he continued to have periodic digestive issues, Ken believed that.

A quarter century had passed since the doctors had cut him open to treat the adhesions. After the surgery, it had taken months to regain enough strength to go back to work in the lab. He was grateful that DuPont had held his job all that time. The last thing he wanted was to have to start over after thirty-nine years at the plant.

Most of the years since then had been good years. He still had stomach issues, but they didn't rule his life. He was able to toss a softball again. He was bowling nearly every week. He had even learned to play golf. He was a leftie, but he had taught himself a right-handed swing.

But the run of good luck didn't last. This time it wasn't Ken but his wife who fell ill. She'd grow pale, suffer fainting spells, and have trouble with her bowels. The doctors couldn't figure out what was wrong. Eventually they performed surgery and removed two-thirds of her small intestine.

The person wheeled out of surgery was, to him, "another woman." She had always been a quiet person. Now she had a raging temper. Ken still loved her, but now he wondered if she still loved him. She wanted a divorce. He couldn't change her mind. They moved out of the house with the three-car garage where they had raised three children. He downsized to a tidy two-bedroom house in Vienna, the quiet suburb of Parkersburg where my grandmother and great-grandparents had once lived. He was fifty-eight now, single, with a lonely old age looming on the horizon.

The one comforting anchor in his life was his work. He now had seniority in the Teflon lab. Promoted to lab analyst, he was a Zone 7 worker, the highest level one could attain in the lab, just below middle management. Finally he was starting to make good money. Now that the kids were out of college, he was able to save for retirement. He planned to retire at sixty-three, when his pension would kick in. After four decades at the company, that should add up to a nice little nest egg.

Never one for sitting around, he had no plans to do it in retirement. He wanted to travel, to visit a foreign country. He thought about teaching math at the local community college. Maybe he'd join a singing group. Or maybe even a singles group.

Then in 2001, about a month after I sent my twelve-pound letter to EPA, Ken's old intestinal pain returned with interest.

In the Teflon lab, he would find himself doubled over, clutching his abdomen, and stutter-stepping to the restroom. It was embarrassing. He hoped that his colleagues didn't notice how often he ran or how quickly. The cramps would come suddenly, unpredictably. As a result, his social life was suffering. He was going out less, avoiding places where there might be a line for the toilet. Because sometimes he didn't make it. When the savage pains in his gut came stabbing back, his life tipped back into limbo. He thought back to the pains that had struck him in the 1970s.

He remembered what the doctor had told him after surgery: "Ken, I got your problems fixed. But I'm worried about your future."

Those words now echoed in his mind. The future was now, and the doctor had been right. Ken had thought the pain unbearable then. Now it felt much, much worse. He started missing work. Some days he could do nothing but writhe in bed, praying for it to stop. The doctors ordered tests that never seemed to clarify anything.

One day he ran to the bathroom and found his shorts filled with blood. By spring, the pain made it impossible to work. He didn't even have the strength to go in and resign in person. He just called and said he wasn't coming back.

Thirty-nine years he had worked there. Nearly four decades. Almost two-thirds of his life. There should have been a retirement party, with cake and balloons and funny toasts. They should have given him a gold watch or something. He should have gone out with a bang, not a fizzle.

After he left the plant, he started to feel a bit better. He took it as an opportunity to drive to Florida to visit his brother, who was struggling with his own failing health. Ken drove through the night to get there. At some point on the drive, his bowels gave way. It was confusing. Scary. Messy. Humiliating.

After he got home, he went to the doctor who had no answers, only some more tests. As he waited for results, the pain spiraled from terrible to "astronomical." It doubled him over and ripped him in half. When his bowels weren't out of control, they were frozen. He'd go four or five days without a movement. It was exhausting, just trying to hang on one more day.

Then the tests came back. "Ken," the doctor said, "I'm afraid we've got a serious problem."

Aggressive rectal carcinoma. Cancer. His history of ulcerative colitis had given him a much higher risk of rectal cancer, which was no longer a risk but a fact.

It's a nasty disease, and Ken would have to bear it alone. At least now he would have a chance to hold DuPont accountable for the hell he was going through, to pay for what they had done.

. . .

As we began to gear up for this new phase of the work that had begun nearly a decade and a half earlier, I was busy putting out fires and managing rumors swirling within the community, mostly thanks to all the new plaintiffs' lawyers who had swooped into town to hunt for the personal injury work. Outside firms had begun to descend upon Parkersburg as soon as the first probable link had been announced in late 2011. Since then, the lawyers had been circling members of the class searching for people like Ken who had suffered from one of the linked diseases.

I found that maddening. It wasn't that we wanted to represent every individual personal-injury case arising from our class members, or even could if we had wanted to. But it was hard not to be galled. Where had these firms been during the fourteen years it had taken to build the case? We had not yet been able to send the class members the formal notice of their rights under the settlement because, under the class settlement agreement, we had to first negotiate its language with DuPont and then wait for the court to approve it. This left a temporary void that other lawyers were happy to take advantage of. Their opportunistic timing was nauseating, but frankly I was more concerned about the problems they'd cause by spreading misinformation. These were firms that had never been involved in PFOA litigation. Did they even understand the most basic facts about the chemical and the case? I highly doubted it. But that didn't stop them from holding public "informational" meetings.

A day before the Science Panel had released its final report, I had driven to Parkersburg to attend one of these meetings. It was hosted by two lawyers I had never met or heard of from firms that had never been involved in a PFOA case.

I had introduced myself and sat in on the meeting. Only a handful of actual residents had shown up, but it was unbelievably frustrating to watch those attorneys standing in front of my team's class members, purporting to "educate" them about PFOA and their rights under our settlement. It was even more maddening when one of the attorneys told the local media this: "People are generally surprised that I'm the only attorney coming to their town and explaining their legal rights."

The court had approved our formal class notice in January 2013, the letters had gone out, and it had turned into an absolute free-for-all. The formal notice had explained that unlike in a normal personal-injury case, where the plaintiff has to fight with the defendant over whether he or she has been exposed to enough of the defendant's chemical to actually cause his or her disease, no class member here would have to wage that battle. Here, now that the probable links with disease had been established, DuPont could not dispute in any of these individual cases that the class member's PFOA exposure was capable of causing the disease—in other words, a plaintiffs' lawyer's dream come true. More than two dozen personal injury lawsuits had been filed against DuPont (none by us). The early-bird lawyers were rushing to sign up as many of our class members as possible as more outside firms swarmed in.

Before the class notice had even gone out, we had been trying to initiate discussions with DuPont to come up with a way to address all the injury claims for the class collectively. But the new legal activity by all the outside firms had complicated those efforts. DuPont suggested another way forward: a multidistrict litigation, or MDL. An MDL is a special legal procedure for streamlining complex cases impacting a large number of people. It's used in litigation involving hundreds or even thousands of individual cases that have many common issues but, unlike class action cases, involve claims where the relief being sought (usually money to compensate for damages) won't be exactly the same for everyone. MDLs have been used in product liability suits in which one drug had purportedly injured thousands of people, airplane crashes in which one crash had resulted in many different injuries to many people, and, notably, in the $20 billion settlement for the *Deepwater Horizon* oil spill. An MDL can also be used when a single event resulted in a wide range of different types of injuries to thousands of different individuals, states, cities, and businesses. The MDL mechanism allows such cases to be consolidated before a single court, which can then try to resolve certain common, core issues in a way that may help simplify and expedite the resolution of all the cases.

From DuPont's perspective, attempting to litigate these cases separately in different jurisdictions, with conflicting rules and timetables, would be a logistical nightmare. Their lawyers and scientists would be forced to appear in different courtrooms at enormous expense for repetitive depositions and trials where separate judges had no qualms about setting schedules incompatible with one another.

This was the rare case where our legal team's interests aligned with DuPont's. Through our class action and its settlement, we already had resolved the core issue of whether exposure to PFOA among the class members could cause serious disease. Although each of the class members diagnosed with one of those diseases might now have unique individual damage claims because of it, there were still a number of core issues related to DuPont's basic liability for having released the PFOA into the environment in the first place that would likely be common to all the cases. Litigating those common issues in thousands of separate cases in different courts—a process that would inevitably be muddied by the free-for-all of plaintiffs' attorneys who were often unfamiliar with the complexities—could take decades and be extraordinarily costly and inefficient for everyone involved. Coordinating all the claims to the extent possible through a single MDL overseen by a single judge in a single court could speed things up greatly for those hurt by exposure to PFOA.

The new DuPont "C8 Personal Injury Litigation" MDL was created in April 2013. All the pending cases were transferred to and consolidated in a single federal court in Columbus, Ohio, which would be presided over by Judge Edmund A. Sargus, Jr. Judge Sargus had been on the bench in Columbus since 1996 and had recently been appointed chief judge of the Southern District of Ohio. Given the many complications of an MDL, the judge's extensive experience would come in handy. We had no idea how big the MDL could grow, as new cases were constantly being filed by multiple attorneys. The cases were split among about a dozen law firms, but most of the plaintiffs wanted our team to represent them.

The MDL process would allow each side to propose a handful of "bellwether cases" for a jury trial, the premise being that jury verdicts on those few presumably representative or typical cases would help inform both sides how the entire pool of cases might play out and thus serve as a benchmark for settlement amounts, if DuPont chose that route. Alternately, DuPont could opt to fight each of the thousands of lawsuits in court individually. Big Tobacco had used that tactic. After the industry's $250 billion settlement with various states over costs related to smoking, there were still tens of thousands of personal-injury cases pending. The tobacco companies had opted to try them individually. The progress had been so slow that personal-injury cases were increasingly turning into wrongful-death cases—filed by survivors of plaintiffs who hadn't lived long enough to see their day in court.

The tobacco companies' combined resources dwarfed even DuPont's. But they had had more than time and money on their side. They'd also had a key argument unavailable to DuPont: they could argue that plaintiffs *had chosen* to smoke, *even when they had known the risks.* Our class members had had no knowledge of the presence *or* risks of the PFOA in their water. They hadn't been given a choice.

In any case, neither Ed and Harry's nor Larry's firm had the capacity to take on the thousands of potential individual trials that might be required if the bellwether process failed and DuPont opted to litigate every case. Ed and Harry's firm was already being overwhelmed with the incredible amount of paperwork and in-person meetings and telephone calls generated by just the initial processing of class members who were asking us to handle their individual injury cases. It was time to call "Pap."

Mike Papantonio was partner at a Florida firm that specialized in mass torts. He ran a legal conference called "Mass Torts Made Perfect," held annually in Las Vegas, where he was treated like a rock star. Many of his fans knew him from *Ring of Fire*, a nationally syndicated radio talk show he cohosted with Robert F. Kennedy, Jr., and other legal experts. Ed knew Pap well. They had worked together on another environmental case against DuPont in West Virginia—the "Spelter smelter case"—which had involved the contamination of the small town of Spelter by a heavy-metals smelter. Ed had assisted Pap as local counsel in the case and had watched Pap work his magic before the jury and recover a huge verdict against the company. Ed had said early on that if it ever became time to take individual PFOA personal-injury trials to court, Pap was the guy.

Shortly after the 2005 settlement, at Ed's suggestion, he, Larry, and I had flown to Pensacola, Florida, so Ed could introduce us to Pap. We were invited onto his eighty-nine-foot yacht, *Nyhaven*, where he hosted parties covered by society reporters who described each guest's attire. Trimmed with inlaid teak and brass, watched over by a sculpture of Triton, it was the only yacht I've ever set foot on (and likely ever will). Pap was a gracious host with a big personality. I could imagine him captivating a jury with his southern charm. He had made a big to-do over toasting our settlement and told us that if links were ever found and folks had to pursue personal-injury trials, that was his specialty and he would be more than happy to help.

Although nearly a decade had now passed, Pap kept his word and agreed to help with the new injury trials. As this was going to play out in the context of an MDL, he supported having an additional firm that he knew well from prior MDL proceedings, the law firm of Douglas & London in Manhattan, help lead the new MDL for all the plaintiff firms. Name partner Mike London was well known in the MDL world and had had unmatched success and experience setting up, managing, and resolving some of the most complex MDL proceedings in US history. His partner, Gary Douglas, was (literally) a rock star. A born-and-bred New Yorker, he played guitar and sang in a band called the Gary Douglas Band. He had started out at a defense firm, as we had, but had formed his own firm that specialized in personal injury and product liability cases. He had an impressive record of big verdicts in complex pharmaceutical cases.

Gary was as northern as Pap was southern. Together they would form a yin-yang trial team that would be a formidable force in the courtroom. The first bellwether case was scheduled for September 2015, more than two years distant. That sounds like a long time, but every week of it would be filled with the arduous process of laying down the procedures for the trials, choosing the plaintiffs, executing discovery, conducting depositions, and, increasingly, responding to DuPont's aggressive attacks on our evidence. They flooded us with more than a hundred motions on that topic alone.

I don't think there was a single day in all that time that I didn't imagine myself finally facing a jury and making it see the facts I'd uncovered over the past fourteen years.

. . .

If I'd thought that with the confirmation of probable links between PFOA and disease finally being known and class members finally being able to pursue their claims for damages I was entering a period of professional satisfaction, I was wrong.

DuPont had brought in an entirely new army of lawyers from the law firm of Squire Patton Boggs, and they were inundating us (and the court) with motions and briefs designed to essentially undo various terms of the original class settlement. This was the big bonus of the MDL setup for DuPont: with the thousands of cases lined up neatly before a single court, if they could succeed in blocking the trial process from even beginning, all those individual suits could instantaneously die.

We couldn't allow that. We were armed to the teeth with scientific evidence and experts—at great expense. DuPont went after them all, filing motion after motion seeking to bar them from testifying, accusing some of them of peddling "junk science."

The court denied every significant motion by DuPont. But for every denial, dozens more motions followed. It was like fighting a Hydra, the multiheaded monster from Greek mythology: every time you chop a head off, two more grow in its place.

Many of the motions sought rulings to prohibit us from showing the jury any of the documents or evidence that DuPont didn't like. These included Bernie Reilly's and John Bowman's emails, documents relating to the Tennants and the Cattle Team, and EPA's lawsuit against DuPont, with its $16.5 million settlement and penalty.

Again, almost all of the motions were eventually unsuccessful. Yet that didn't stop the paper onslaught. When the court ruled against

DuPont, their lawyers would simply repeat and reassert the arguments in slightly different form in an even greater number of motions, over and over again. Death by a thousand paper cuts.

There wasn't time for everyone working for the plaintiffs in the new MDL at all the additional firms to read the millions of pages of documents I had read or learn all the facts I had memorized. I quickly gave up trying to bring everyone up to speed on the enormous body of evidence and the subtle nuances of the science. It was often faster to do the work myself, even if that meant night after night spent researching and writing until 3:00 a.m. or later. Sarah kept begging me to slow down, worried that the renewed stress would trigger a neurological episode. She rarely complained about my travel or work commitments, and although I managed to be physically present for my family most evenings and most weekends, I was rarely emotionally there. My head was in these cases. I doubted the boys, who were now all teenagers, needed much more than my physical presence at home and at their soccer games, art shows, and school functions. But I knew Sarah needed my head in the game at home. That became even more difficult for me because I had been appointed by the court as one of the co–lead counsel for all the plaintiffs in the MDL and had enormous responsibilities for thousands of cases and claims. And aside from the sporadic mild symptoms—the twitching and jerking in my neck, arm, legs, and eyelid—I had been fine for a long time. So I told Sarah not to worry, that I would be okay. The truth was, I was worried, too. But I was so close to the end. Too close to stop.

• • •

Before we got to trial, the fiercest and most significant battle was over general causation, something DuPont had already agreed would be settled by the unbiased Science Panel they had helped create. This was the basic question of whether PFOA was capable of making class members sick. When negotiating the 2004 class settlement, I had taken great pains to be sure it was clear that DuPont had agreed not to contest general causation for any disease the Science Panel linked to PFOA. In other words, they would not dispute the fact that PFOA was capable of causing those particular diseases at the exposure level required to be a class member. According to the settlement, as long as someone met the definition of being a class member (exposed to 0.05 ppb or more PFOA in his or her drinking water for at least one year), he or she was now understood to have been exposed to enough PFOA to be capable of causing any of the six linked diseases, period. That was general causation. And that was the whole point of having both sides agree to the Science Panel—to resolve the basic general causation issue for every single class member. DuPont could still contest *specific* causation. That referred to whether the PFOA not only had the ability to cause a specific disease but was a substantially contributing factor to actually causing the disease in a particular individual, as opposed to other possible causes, such as smoking, obesity, and genetic predisposition.

I could tell from the briefs and arguments submitted in the new MDL that DuPont was laying the groundwork to attack the cornerstone of our settlement agreement: they planned to challenge the Science Panel's established links between PFOA and the six diseases for all class members, despite the now decade-old written settlement agreement promising precisely not to make any such argument. The fiasco when the company had prevented Brookmar from running the medical-monitoring program had taught me not to assume that they would abide by a written agreement. This time, I wasn't going to hold off until it was too late to demand compliance with the terms of a deal. The last thing I wanted to do was wait until trial to fight the battle over general causation all over again. So I decided to force the issue up front with an unusual legal tactic, filing a motion asking the judge to rule that this one issue was already settled and therefore not subject to argument in the upcoming trials. Some of the other plaintiffs' lawyers in the MDL who had not been around for the settlement talks viewed it as a risky move—what if we lost and the court allowed DuPont to essentially rewrite the settlement terms on general causation? But I had been there in Boston when we had negotiated the deal. I had carefully crafted the language of the formal written agreement to prevent DuPont from disputing exactly what I now suspected they were going to try to dispute. I had faith in the settlement, enough to override the concerns of the other plaintiffs' lawyers and ask the court for early summary judgment on this one issue.

. . .

The hearing on my request convened in Columbus on November 13, 2014. It got off to an amusing start when DuPont's lead counsel, Damond Mace, took off his overcoat and revealed that he had forgotten his suit coat. For a corporate lawyer, that was a little like a knight showing up for a joust and realizing he had forgotten to put on his armor.

Mace apologized for his breach of decorum.

"No apology necessary," Judge Sargus said. "But you have to admit, it is a bit humorous, maybe not for you but for the rest of us."

In fact, I did find it funny. But more than that, I hoped it was a sign. I knew that the fate of the entire MDL could be wrapped up in this one seemingly limited issue. If DuPont could go back to trying to pick at the very basis of our settlement, resolving once and for all the issue of whether PFOA indeed had serious health impacts for the people who had been drinking water contaminated by the Washington Works plant, we would lose a decade of work and basically have to start again from the beginning. So despite my faith in the strength of the settlement, my stomach churned with anxiety. If this hearing went the wrong way, we were, to put it crudely, screwed.

I pushed myself up from the plaintiffs' table, buttoned my suit coat ostentatiously for both the judge's and Mace's benefit, and placed my legal pad and pen in front of me on the podium.

I cleared my throat. "Your Honor realizes this is an extremely unique case, a very unusual situation," I began. "We're governed by this very unique contract with one of the world's most sophisticated companies that agreed to an unprecedented agreement with seventy thousand people, that we're going to look at your exposures, your actual specific exposure level . . . and we're going to agree on a panel that's going to tell us what diseases can actually be caused from PFOA at these levels. And if we do find that something is caused, we, DuPont, are telling this community, we will not dispute that PFOA was a cause in these cases going forward. . . .

"Granted, it's unusual," I continued, "but we have eighty thousand people that agreed to this, waited years for this, came in, had their blood tested, went through all kinds of medical procedures for this. . . . DuPont should be required to abide by this agreement."

Judge Sargus said, "But this is where we start to get the grind. I've gone back and read the scientific findings, and they use a lot of data. They look at the people who have been consuming the water, for example. But they look at other matters as well. . . . They didn't specifically say dosage amount, though, I don't believe. Did they?"

Here was an opportunity to shine a spotlight on the crucial point. "Exactly, Your Honor, and here's why: because the charge up front already told them what dosage they're looking at. . . . The parties already have the defined exposure in mind. They wanted to know what diseases would be caused in this group." I was reminding the judge that we had specifically defined the class to be studied by the Science Panel as everyone with a specific "dose" of PFOA—at least 0.05 ppb in his or her water for at least one year. The panel's task had been to confirm whether there was a probable link among that group of people, which had been defined by its dose of PFOA up front.

Judge Sargus nodded. "Right," he said.

I was hoping the court could see that the basic deal we had struck all those years ago had settled the issue. The panel hadn't spent seven years investigating whether PFOA—in some undefined concentration—was likely to cause disease in some humans somewhere. The Science Panel had looked to see if PFOA was capable of causing disease among this particular group of people exposed to 0.05 ppb or more of the chemical in their drinking water for one year or more. And it had concluded that yes, it could—a conclusion that DuPont had agreed under a written settlement agreement—a legal contract—not to dispute.

Once the Science Panel had confirmed that PFOA had a probable link to a specific disease, I explained, it was a given that everyone in the class had been exposed to enough PFOA for it to have caused that disease in them. No one in the class had been exposed to *too little* PFOA to qualify.

Judge Sargus nodded again. "Right."

"The links were found in the entire group, Your Honor."

After a bit more back-and-forth, the judge said, "Well, let me back up here. . . . With smoking, the longer you smoke, the greater your risk. I don't know if that's true with regard to PFOA. . . . In other words, will five years of drinking the water versus one year make a difference?"

I needed to make clear that this was not the typical tort case where the standard fights between experts over dose and causation would rage on; that had been the whole point of our settlement: to avoid those types of long-drawn-out legal battles. This was at its core an issue that was governed by the simple terms of a contract written a decade before—a deal. Once the Science Panel had confirmed the probable links, they applied to all class members equally, and DuPont had agreed to abide by its findings. End of discussion on any issue of general causation for any class member on any linked disease. Period.

"Your Honor, we believe that issue would be off the table."

I was relieved when I heard Judge Sargus respond, "That's general causation."

By the time I collected my legal pad and sat back down, I was feeling pretty good. Then Mace approached the podium, coatless, and I had to remind myself to keep a poker face. But Mace is an excellent lawyer, just like the rest of DuPont's legal team, and he could actually count the number of angels who fit on the head of a pin. Or, more relevantly in this case, he could try to pry open gaps in an ironclad agreement designed to be gapless.

"As the court knows," Mace said, "in the settlement agreement at line—let me grab my copy, Your Honor—it's provision 1.49, it says, 'Probable link shall mean that based upon the weight of the available scientific evidence, it is more likely than not that there is a link between exposure to PFOA and a particular human disease *among* class members.' So the key term, as the court will recognize, is the 'among.'"

I couldn't believe he was really going there. But he was.

"If you look that up, Your Honor, in *Webster's Dictionary*, one of the silly examples they use is there were ducks scattered among the geese.

That's the sense in which it's used. It doesn't mean that all the animals were geese.

"In here the probable-link language does not say 'all.' It does not say they are finding this association in all the class members. If the parties had intended that, they could have used that simple language. But that's not what it says. And the plaintiffs can't add words or change words that are there."

Now I was angry. A decade after making a clear commitment to abide by the decision of the Science Panel for all class members, DuPont was coming up with ducks among geese? I was about to get heated, but the judge saved me the trouble.

Judge Sargus stopped him midsentence. "The problem is, let's go back to the language, that is really key here. I read, 'defendant will not contest the issue of general causation between PFOA and any human disease.' There's no limitation as to what members of the class this is being applied to, is there?"

Mace paused. "Excuse me, Your Honor?"

"I'm reading," the judge responded.

"Yes, sir," Mace said, his cheeks reddening slightly.

Judge Sargus continued, "My point is, the conceding of general causation, once the Science Panel made their decision, it's not limited there at all. It means—I read that to mean 'to any class members.'"

Mace resisted. "No, Your Honor," he began. As I listened to what followed, I couldn't believe what I was hearing. He was going for semantics, the same way Bill Clinton had when he had famously told a grand jury considering whether he had indeed had sex with "that woman," "It depends on what the meaning of 'is' is."

"First of all," Mace said, "it's that it's capable of causing disease. And then, the only argument they have to try to tie this into all class members is this 'among class members.' But it doesn't say 'all class members.' It says 'among.' And what you have to keep in mind, Your Honor, factually—"

The judge cut him off. "I don't—I'm no *Webster's Dictionary*. But 'among' doesn't seem to be a limiting qualifier."

• • •

The court didn't rule immediately that day. But a month later, when the ruling popped into my computer, I had to smile when I got to the part where the court flatly called DuPont's argument "untenable," rejecting any claim that they could question the logic of the Science Panel's probable link decisions due to "the unambiguous language" and "the clear contractual language."

It was a satisfying and critical win, but even after getting slapped down in such an explicit way, DuPont *still* continued submitting expert reports disputing general causation, and each time we had to move to get the experts or their opinions thrown out. This continued all the way to the day of the first trial. But each time, the court held them to the terms of the settlement. After all, we had upheld our end of the bargain. Although the Science Panel had confirmed links with six diseases, it also had said that it was unable to confirm any such link with more than fifty other specific diseases. That didn't mean there was no association with those other diseases, only that the panel had been unable to confirm them as probable links among the class members, so those diseases were off-limits for our class. DuPont had readily accepted those "no-link" findings as applying to all class members (regardless of any individual dose levels) and had readily accepted the class members' releases and dismissals of those claims for the benefit of DuPont. The company could not have it both ways.

Through all that, to the outside world, DuPont kept a brave face, refusing to budge or show the slightest bit of contrition. "We are confident that we acted responsibly at all times," a spokesperson told the *Parkersburg News and Sentinel*. "We will continue to vigorously defend these cases."

But inside DuPont it was another story and had been for some time. Way back in 1992, DuPont's lawyers had predicted that "toxicity issues associated with C-8" could turn it into the number-one DuPont tort issue. As Bernie Reilly had acknowledged in an email to his son, juries tended to be more sympathetic to injured humans than to giant corporations. "[M]ost simply do not believe how big and bad we would look before a jury." And as John Bowman had predicted, this was when the company would "spend millions to defend these lawsuits and have the additional threat of punitive damages hanging over our head."

DuPont made a motion to the court to prohibit us from seeking punitive damages in our bellwether trials, arguing that no rational juror could find any basis for punishment in the underlying facts.

I jumped at the opportunity to work on our response. I pulled together and attached for the court a sampling of all the evidence we had collected over the years. DuPont's motion was flatly rejected by the court: "A reasonable jury could find the evidence shows that DuPont knew that PFOA was harmful, that it purposefully manipulated or used inadequate scientific studies to support its position, and/or that it provided false information to the public about the dangers of PFOA."

With that, after eleven years, we were now for the first time on the verge of discovering exactly what a reasonable jury would make of all this.

33

THE TRIAL

September 15, 2015
US District Court, Columbus, Ohio

On a late-summer morning I walked up the steps of the eighty-four-year-old federal courthouse in Columbus, Ohio. Built in the 1930s, the Joseph P. Kinneary U.S. Courthouse is a monolithic Neoclassical building that takes up an entire city block along the Scioto River, which flows into the Ohio. With thick sandstone walls and towering columns, it looks like a modern-day Parthenon. At the top of one facade, the stone is carved into an allegorical scene depicting justice flanked by industry and agriculture. Looking closely at the details, I noticed a farmer leading two cows. I took that as a good omen.

Every step along the marble floor of the hall took me closer to the place and time I had been working toward for more than sixteen years. As I put one foot in front of the other, I felt suddenly anxious. What if I had an episode and my leg gave out, as it sometimes did? What if I stumbled or began to jerk as I entered the courtroom? I fought those fears down by trying to focus on the solid reality in front of me: our team had built a very strong case. But even without my physical concerns, I would still have had to shake off the courtroom jitters.

The trial had officially started the previous day with the swearing in of the jury. But this marked the commencement of the real courtroom action. Mike Papantonio and Gary Douglas would make the opening statements. My first job at the trial was to give the opening of the opening. I would be setting the stage for the rest of the trial. I would provide an overview of the prior sixteen years, then introduce Pap and Gary to the jury and hand over the reins to my superbly capable lieutenants.

By a quarter to nine, the courtroom was packed. The gallery hummed with restless energy as reporters reached for their notebooks, attorneys thumbed messages on their smartphones, and spectators chatted expectantly. The murmur died promptly at 9:00 a.m. as the jurors filed in silently and took their seats in the jury box on the right side of the courtroom.

"All rise," the bailiff said.

We stood as Judge Sargus entered the courtroom and took his seat at the bench.

"The United States District Court for the Southern District of Ohio is now in session, the Honorable Judge Edmund Sargus presiding."

"You may be seated," Judge Sargus said. The shuffle of a hundred people sitting down at once gave way to an electric silence.

"Your honor," the bailiff said, "today's case is Carla Marie Bartlett et al. versus E. I. du Pont de Nemours and Company."

"Counsel," the judge said, "you may proceed."

Carla Marie Bartlett was one of the cases that DuPont had nominated to be a bellwether case. Carla Bartlett was a fifty-nine-year-old grandmother who had been diagnosed with kidney cancer after drinking PFOA-contaminated water for seventeen years. We presumed that she had been chosen because, in the large array of several thousand cases, hers was representative of those with lesser potential damages; her disease had been successfully treated at minimal cost to her. And there was another reason for DuPont selecting this case that would become clear in cross-examination: she was representative of the group of plaintiffs for which DuPont thought it could try to blame the disease on something other than PFOA. In other words, this was arguably a "best case" situation for DuPont, and they wanted to see how that would play out in front of a jury.

I took a deep breath, rose from the plaintiffs' table, and stepped to the podium in the center of the room.

"May it please the court, ladies and gentlemen: good morning, my name is Rob Bilott."

As soon as I heard my words echo through the courtroom, I relaxed. Finally, as I had imagined a thousand times, I was standing in front of a jury to tell the story of PFOA. I took a deep breath and met each pair of eyes. Three women, four men. The jurors blinked back at me, their faces neutral and attentive. They would be the first ordinary citizens ever to hear the whole story. To make them understand, we would have to take them back—way back—to decisions made long before Mrs. Bartlett had ever gotten cancer. To the beginning before the beginning.

"This is a case that spans a wide range of years, goes back many decades," I said.

I had hardly slept, but I was fully alert. I had scribbled a few bullet points on a scrap of yellow legal paper. I pulled the page from the pocket of my very best suit, unfolded it, and smoothed it out on the podium. But I didn't look at it. I didn't need to.

"You're going to be seeing a lot of documents going back quite a bit of time."

By now I knew them all by heart. I could recite every date of every key document we would use as evidence. I could quote them verbatim. I had read and reread them so many times that they were burned into my memory.

Each juror had been given a pen and a legal pad. They were encouraged to write down the identifying number of any document they wished to study up close. At the end of the case, each of those documents would be delivered back to the jury room. I hoped—believed as I always and so far naively had—that if people could only read them, they would see and understand what I had been seeing all these years.

This case was focused on whether or not PFOA had caused

Mrs. Bartlett's cancer. But I wanted the jury to understand where Mrs. Bartlett fit into the long, twisting saga of PFOA. I wanted them to see the big picture.

"We wanted to give you a little bit of context on what you're going to see," I said, "and how we got where we're at today."

What a shame that Earl couldn't be here now. Even in that moment, standing before the jury with the judge and the gallery watching, I thought with regret that I had never had the chance to say good-bye. The only way I could find peace with that was to hold DuPont accountable. That's what Earl had wanted most. To do that, we would have to win this case.

I had to finish what he had started.

"This dates back to a situation that occurred in the late 1990s," I told the jury, "with a family with the last name of Tennant . . ."

. . .

After setting the stage with my five-minute prologue, I turned the floor over to Pap and Gary for the remainder of our opening statements. Pap went first. I watched him stride before the jury with measured confidence, wielding words and emotions like weapons. He had enough personality to fill *two* courtrooms.

"Good morning, ladies and gentlemen," he said warmly. "I'm from north Florida. I'm going to speak as clearly as I can, but there is a difference between me and Gary Douglas. He speaks a little faster."

Gary, the New Yorker, smiled back at him from the plaintiffs' table, where we sat on the right side of the courtroom. I sat between Gary and Carla Bartlett, with the jury off to the right. In that courtroom, the plaintiffs sat closest to the jury box, where the jurors could watch them up close. From time to time Mrs. Bartlett would lean over to me and whisper a question. Most of the time she sat quietly, trying not to fidget when the jury studied her.

As we had prepared for trial and I had gotten to know Mrs. Bartlett, I couldn't help but like her. Soft-spoken and warm, with blue eyes and chin-length fine blond hair, she walked with a slight limp, leaning on a cane. When we spoke, she had a way of putting her hand gently on my arm, which melted my heart. She had a nineteen-month-old blue-eyed grandson whom she clearly adored.

Carla Bartlett was born in Parkersburg and raised in rural Ohio, just across the river from Washington Works, in a village with a population of about twenty. The youngest of five, she could remember the days when water for drinking and cooking had been collected from a natural spring. Their father had eventually dug them a well. As a young adult, she began drinking Tuppers Plains public water. Her father spent most nights on the road, building highway guardrails. He came home to his family on the weekends.

During her high school years, she had worked after-school and summer jobs on nearby farms, helping out in the fields of corn and tomatoes. In her twenties, she moved out of her parents' house and into a trailer on their land. She worked for a family-owned real estate business, keeping the books. In the evenings she worked a second job at a convenience store. That's where she met Jon Bartlett, her future husband.

On their wedding day, the father of the bride was too sick with the flu to walk his daughter down the aisle. So Carla walked on the arm of a little boy, the groom's young son from a previous marriage. He barely came up to her elbow. The boy delivered the bride to the altar, where the minister asked him a question: "Jon Michael, do you accept Carla as your stepmother?"

"Yes," he said. "I do."

In November 1995, Carla and Jon gave the boy a little brother named Alex. Five or six weeks later, on New Year's Eve, Carla doubled over in pain. Friends looked after the kids while Jon rushed her to the emergency room. It was her gallbladder, doctors said. But there was something else: the CAT scan also showed a spot on her kidney.

She would get the chance to tell the rest of the story to the jury—soon. First we had to lead the jury through the science, the documents, and the management decisions. Our expert witnesses would testify first, to lay the framework and the context. Then Carla Marie Bartlett would step up onto the witness stand and testify from the heart of it. It was up to her to explain what it was like to be an involuntary participant in DuPont's vast human health experiment.

. . .

Outside the courtroom, the media was buzzing about the case.

Just months before the trial began, DuPont had "spun off" the Teflon part of their business into a new company called the Chemours Company. It had happened in the context of a stockholder revolt against alleged inefficiency—the spin-off had come with "streamlining" (read "layoffs")— but as soon as I saw the news, I was concerned that it was an attempt to avoid liability. Something similar had occurred a few years earlier when the energy company Kerr-McGee, made infamous by allegations that it had purposely or negligently exposed the labor union activist Karen Silkwood to plutonium, had spun off a company called Tronox, which had eventually gone bankrupt under the weight of huge liabilities for wide-scale environmental damage caused by the parent company. Though a court eventually had ruled that the spin-off had been a fraudulent transaction and the parent company had to make good on the liabilities, it had taken years of nightmarish legal wrangling. I wanted no part of that kind of situation. In our case, we had eventually forced DuPont to affirm that they would remain responsible for any judgments resulting from our trials, but as the trial continued, unnerving speculation swirled through the financial market: did Chemours have deep enough pockets to afford the potential liabilities?

Wall Street was not optimistic. One litigation analyst estimated that Chemours' liability could total up to $498 million. Another analyst at a different firm said that any amount above $500 million could drive Chemours into bankruptcy. Gauging from the stock price, investors were equally pessimistic. The Chemours share price had plummeted 57 percent since June.

Many people were drawing comparisons with the Big Tobacco litigation of the 1990s, which was still dragging on in US courtrooms. As in our case, an initial class action had resulted in the issue of general causation being resolved. Because of that earlier class action, class members no longer had to prove that, "generally" speaking, smoking could cause cancer. That was a given. But they still would have to prove specific causation—not only that smoking *could* potentially cause cancer but that it, rather than something else, *had* caused cancer in the individual plaintiff.

. . .

"This case breaks down pretty easily," Pap told the jury. "What did Du-Pont know about how dangerous this product was, and how soon did they know it?"

He then proceeded to outline the case we would show to the jury over the next few weeks to prove that DuPont should be held legally responsible for Mrs. Bartlett's injuries. He highlighted dozens of the documents I had unearthed over all of the prior years and explained how they all fit together like pieces of a giant puzzle that the jury would get to see assembled right before their eyes.

As I watched and listened to Pap lay the story out before the jury, I found myself struggling not to cry, shocked by a powerful welling-up of some combination of joy and grief and terrified that I would lose it in front of the jury and the entire courtroom. Then the realization slowly sank in: I was finally seeing everything I had worked so hard to uncover and put together. All those hours spent digging through documents, all those years of waiting, all those motions, briefs, arguments fighting over every imaginable legal issue and piece of evidence, all those people depending on me to make things right for them: this was it. It all boiled down to the story that the jury was now hearing from Pap. Judgment day had arrived at last.

As I struggled to maintain my composure, Pap finished his overview of DuPont's culpability. Then Gary got up to take his place in front of the jury. He had a much more personal story to tell: how Mrs. Bartlett had suffered as a consequence of DuPont's actions and omissions that Pap had just explained. As I listened to Gary explain just how terrible Mrs. Bartlett's suffering had been, I thought of the thousands and thousands of other people who needed our team to get this done and the seemingly endless journey I'd been on that now, finally, had an end in sight. I could only pray that end would be justice.

. . .

After the openings were over, the real work of the trial began. For the next three weeks, we called experts and other witnesses to the stand to explain the story and to show how all the evidence proved that DuPont should be held responsible for the PFOA disaster. To do that, we had to prove four basic things:

Number one: Duty. Did DuPont have a duty of reasonable care not to cause injury to Mrs. Bartlett and the other members of our class?

Though the answer might seem obvious, the question invoked a fierce battle.

We argued that DuPont had a duty to avoid polluting public drinking water with a chemical they knew to be toxic. Once they had realized the water was contaminated, they had had a duty to disclose it. "One thing they could have done is real simple," Pap said. "Warn people. Give them a choice."

DuPont claimed that they "never had any knowledge or expectation . . . that there was *any* likelihood of *any* harm to community members at the relatively low PFOA levels found outside the plant." The company argued that they "never knew, nor should have known" that PFOA exposure was hazardous or potentially harmful.

That was the basis of DuPont's defense, the old standby: We didn't know!

One of the elements of duty was foreseeable harm. Could a reasonable corporation in the same position foresee the potential for harm to the plaintiff?

Damond Mace, DuPont's lead trial lawyer, told the jury, "As you look at the decisions that were made by individuals at DuPont in certain years, you need to avoid the 20/20 hindsight. You need to focus on what was known by those individuals at the time the decisions were made."

We pointed out that by 1984, at the latest, DuPont had already known about PFOA's potential for harm and had still made certain decisions: to continue using and emitting PFOA; to withhold toxicity data from EPA; and to conceal the truth about water pollution from the government and the public. That was twenty-two years after the 1962 rat study that had showed enlargement of the kidneys and testes; five years after the 1978 primate study in which all the high-dose monkeys had died and another rat study had found enlargement of the kidneys and testes; two years after DuPont's medical director had warned of the "great potential for current or future exposure of the local community from emissions leaving the plant."

According to DuPont, all of that was not enough to foresee potential harm to people in the local community, such as Mrs. Bartlett. Damond Mace accused our witnesses of being "Monday-morning quarterbacks."

Number two: Breach. Did DuPont act as a reasonable corporation would have acted in similar circumstances?

Our expert witnesses testified what a reasonable corporation would have known and would have done with the information available to DuPont at the time. They explained that DuPont's breach went beyond egregious omission. Not only had they failed to warn the public of the contamination, they had taken actions to conceal it.

"After they knew about toxicity, after they knew about something called biopersistence, after they knew about cancer in laboratories," Pap told the jury, "they told the public, *Don't worry about it. There is no problem here.*"

DuPont stayed on point, repeating the mantra they had consistently told for a generation; there was no reason for the company to have fore-seen any potential harm to people drinking PFOA at the "trace" levels that had been found in the community's water at the time.

Number three: Damage. Did the plaintiff suffer actual injury or damage?

We'd let Mrs. Bartlett herself take the witness stand to answer that question.

Number four: Proximate cause. Did DuPont's actions (or omissions) result in Mrs. Bartlett's cancer?

We reminded the jury that in this case, the issue of PFOA's ability to cause the type of cancer Mrs. Bartlett had was already resolved by a scientific panel (our Science Panel and its probable-link findings) and that was not an issue in dispute for it to resolve here. The only disputed issue on causation was specific causation: whether the PFOA in her drinking water (and not something else) had been a substantial factor in causing Mrs. Bartlett's cancer. And on that point, there was intense disagreement. "Just because PFOA is capable of causing kidney cancer," Damond Mace said, "doesn't mean it caused Mrs. Bartlett's cancer."

. . .

After weeks of testimony from our witnesses and cross-examinations by DuPont, it was finally Carla Bartlett's turn to take the witness stand.

She looked nervous, but Gary Douglas led her gently through her testimony. After answering a few questions about her childhood, she seemed to relax. I watched the jury closely as she began to speak.

Mrs. Bartlett explained that after the doctors had found a spot on her kidney, they'd told her to come back for a follow-up. At the time, she was more worried about her baby boy, who had been born with a double hernia. Still just a few months old, Alex needed his own surgery. Some doctors said it was urgent. Others told her he was a little too young.

When he was old enough, Alex had surgery. Things seemed to be fine, but then Carla's mother died suddenly and unexpectedly. She found herself dealing with an infant, a young boy, and grief. "I just kind of pushed myself to the background," she said. Her husband, Jon, was a truck driver who traveled all week and came home to his family on the weekends. So when her doctor's office called her in for a CAT scan, she tried to put it off. "I don't have time," she said. "I've got a baby. There's just so much going on."

She'd never forget the reply: "You don't have time *not* to do it."

After the CAT scan, the doctor's office called again. "Mrs. Bartlett," the voice on the phone said, "the spot on your kidney . . . they're 98 percent sure it has turned to cancer."

I could see she was getting emotional. Gary noticed it, too, and wanted the jury to understand.

"Mrs. Bartlett, what was going through your mind then?"

"Utter terror. I was so worried, because at that point I realized that, you know, I—I could very well have cancer and I may not be able to be there for my family."

"And what did you do after you received that phone call?"

"I just started shaking and I was really, really scared. I called my husband, Jon, and told him that I was going to have to come see a surgeon, a specialist, at James Cancer Hospital."

"How old were you at the time?

"I was forty-one."

"How did it feel to hear those words?"

"When you hear the word 'cancer,' it scares you anyway. But then when you hear the word 'cancer' associated with yourself, it makes it even worse. And all I could think of was . . . not being there for my family."

The spot on her kidney had grown into a tumor the size of a grape. "Renal cell carcinoma," the doctors said. They'd caught it early—at stage 1—but removing it would require an operation her doctor described as "cutting a patient in half."

Gary led her through the day that her husband, Jon, had driven her two and a half hours to a hospital in Columbus for the surgery. The thoughts that had entered her mind on that drive. The silence that had filled the car.

"I was—I was scared to death," she said. "I just kept trying not to cry because my husband—I didn't want to upset him more than he already was."

As they wheeled her into the operating room, Jon had walked beside the gurney until they wouldn't let him go any farther.

"At that point in time, you do the 'I love you.' He said, 'Everything is going to be fine.' I said, 'I love you, too.' I said, 'Make sure, if something happens, that my boys know I love them.'"

The deputy clerk handed her a tissue.

"I remember looking up at the lights, and they were saying, 'Now we're going to put you to sleep. And when you wake up, it will be all over.' So that was the last thing I remember, was looking up at the lights and saying: *Please, God, let me be okay.*"

"What is your first recollection of waking up following the surgery, Mrs. Bartlett?"

"Oh, pain," she said. "It was terrible pain. I'd never had as much pain in my whole life."

She described the dry heaves, the staples, the drainage tube. How her own baby had turned his head away and buried his face the first time he had seen her. That was a different kind of pain.

Pain was a new constant in her life. She had spent months sleeping in a recliner because it hurt too much to lie flat. One day she had stood in front of a mirror, lifted her shirt, and thought, *Oh, my God.*

Gary decided against having her show the jury the scar. "Could you just describe what that scar looks like?" he said.

She traced her finger across her body from her stomach to her back. "It's very big, and it's very ugly."

The wound had eventually healed. In time, the pain had gone away. But the fear would always linger.

"This was cancer. This could come back. And the scar reminds me of it every day."

"Is your fear of cancer coming back worse today than it was back then?"

"I thought everything was going to be fine. . . . Now I'm not so sure."

"Why is that?"

"Because of the level of PFOA that I've discovered in my body."

In 2005, the year Mrs. Bartlett walked into a Brookmar trailer for a blood test, she tested at 19.5 ppb. That meant she had more PFOA in her blood than 99.6 percent of the US population.

As I sat at the counsel table watching Carla on the stand, feelings of intense sadness swirled inside me. Look what DuPont has done to this woman, I thought. Look what they did to Earl, to Sandy—to everyone. My sadness was soon replaced by anger as the cross-examination began.

For the cross, DuPont sent in a new face for the jury. Stephanie Niehaus was a fortysomething female partner from the New York office of the Squire Patton Boggs firm. DuPont had been sending her into our contentious pretrial conferences to argue over every piece of evidence, every legal precedent, every detail on virtually every issue that had come up.

But thus far in trial, she had not spoken in front of the jury. She had been sitting quietly at the counsel table with the DuPont team whenever the jury was present. Now she rose to begin Mrs. Bartlett's cross-examination.

We already knew from pretrial motions that DuPont's main defense on specific causation was that "something else" had caused Mrs. Bartlett's cancer. What was that "something else"? Obesity. In so many words, DuPont would argue that Mrs. Bartlett had gotten kidney cancer not because of the massive amounts of PFOA that had poisoned her water for all those years but because she was too fat.

I guess DuPont actually thought the argument might fly better if it came out of the mouth of a younger, thinner female attorney. I braced myself. This wasn't going to be pretty.

"Would it be fair to say that, while your weight has fluctuated, 250 would be about an average weight for you in your adult life?"

"Yes."

"Okay. And your height is five-foot-six?"

"I think, approximately, yes."

"Okay. And over time, is it fair to say that your physicians have counseled you about losing weight or reducing your body mass index?"

"I'm sure—probably—yes."

Her strategy was transparent and hard to watch.

Mrs. Bartlett blinked back tears, bearing the indignity with all the grace she could muster. Watching, I felt my rising anger turning to rage.

These people were the terrible price of corporate malfeasance. Such costs were not accounted for on any balance sheet; Mrs. Bartlett's cancer was not part of the corporate math.

• • •

With the help of our experts, we had taken the jury on a three-week tour of science, industry, and medicine. Through the memos, emails, and meeting notes, we had given its members a rare peek under the hood of a giant chemical corporation. Through all the internal studies, we had toured the toxicology labs of Haskell.

Through the videotaped deposition testimony of corporate medical director Dr. Bruce Karrh, epidemiologist Dr. William Fayerweather, and the statements of attorneys Bernie Reilly and John Bowman, the jury members had heard employees across the company trying to get DuPont's management to do the right thing. And we had shown them indisputable evidence—from the company's own files—that the corporate decision makers had ignored them.

We had presented the jury with hundreds of documents that spoke for themselves, a collection of evidence DuPont called a "parade of horribles." The jurors had seen the "connect the dots" presentation and its unequivocal objective: "Is there a strategy we can use to minimize the amount of information being disseminated?" They had read Bernie Reilly's emails to his son, describing the escalating situation in the most candid layman's terms. They knew that DuPont's own legal team feared that the company could be facing massive punitive damages. I felt confident that we had made a strong case. We had addressed the four basic elements of negligence. We had spelled out the details of who, what, when, where, and how the whole thing had come about. But there was one other question that could hammer the last nail into the coffin: *Why?*

At the heart of this question was a terrible truth. "They had the alternative," Pap told the jury. "They knew they could clearly use it. But they knew that the problem was the cost."

Back in the 1980s, as evidence of the PFOA contamination of drinking water and human blood was mounting, DuPont had created a special internal team to investigate the problems and evaluate solutions. The most obvious had been to search for a PFOA substitute. The manufacturing lines had conducted production tests to evaluate performance. Haskell had tested for toxicity. The business side had analyzed the costs.

And the company thought they had found a viable alternative—as early as 1983.

It was called telomer B sulfonic acid, or TBSA. Chemically different from PFOA but with similar surfactant properties, TBSA reportedly decomposed "to presumed innocent by-product" when heat treated in industrial ovens. Its ability to accumulate in blood was similar to that of PFOA, but it appeared to be less toxic. However, the team noted that more toxicity tests and blood data would be needed. The problem was, TBSA was more effective for some products than others.

The other solution involved the destruction or treatment of PFOA. The original manufacturer, 3M, had provided DuPont explicit information on the disposal of its product: incinerate it or dispose of it in a facility designed to accept chemical waste. Another option was "scrubbing and recovery"—a method that could capture PFOA for potential reuse.

The point was that DuPont had had options; there were things they could have done to address the safety risks of PFOA. These available options had been promising, but they would have cost money. At Washington Works, thermal destruction would have required an initial investment of $1 million, followed by $1 million in annual operating costs. Implementation would have taken one and a half to two and a half years. Scrubbing and recovery would have cost more: a $3.5 million up-front cost, a $1.5 million development cost, and $2.5 million a year. It would also have required more time: four to five years. But if the recovered PFOA had been reusable, there would have been a potential to break even.

We asked the jury to put those costs in perspective. The DuPont documents indicated that Washington Works' annual operating budget was approximately $650 million. In that light, the cost of any of those treatment options seemed trivial.

We hoped that this reality would resonate: DuPont had known of ways of getting rid of PFOA before it entered the environment.

It seemed like a no-brainer, the kind of action a reasonable company would make with the information available at the time. Especially considering the stakes: the health of an entire community, including DuPont's own workers, versus a relatively low additional cost.

So which option had DuPont chosen? Thermal destruction or scrubbing and recovery?

Neither.

With a flourish, Pap pulled out the poorly spelled PFOA executive meeting notes from 1984—the document I had been waiting for years to show to a jury. The one that said that by eliminating PFOA, the company would "essentially put the long term viability of the bussiness [*sic*] segment on the line."

"They didn't even *have* to use PFOA," he told the jury. "You'll see the only reason they used it is because they wanted to save money. It was about trying to cut overhead, trying to increase profits."

Had DuPont followed through on the alternatives available to them, we might not have been gathered in this courtroom, pointing fingers and measuring scars.

The whole thing could have been prevented. But DuPont had found the costs too high. So now people like Mrs. Bartlett were paying the price.

"It's called externalizing costs," Pap said. "And you can make a lot of money doing it until you end up in a courtroom like this."

· · ·

In closing statements, DuPont's lawyer, Damond Mace, called our case "a house of cards." Like Pap and Gary, he exuded confidence as he delivered his final monologue.

"Three weeks ago I stood before you and I told you I'd prove three things for you. One, that no employee of DuPont thought that their actions were likely to cause any harm to Mrs. Bartlett or anybody in the community. Two, that DuPont has no liability here. It did not breach any legal duty to Mrs. Bartlett. And three, that Mrs. Bartlett's kidney cancer is readily explained by other things and not caused by any conduct of any DuPont employee.

"Persistence does not equal harm," he said, a theme of his case. He reiterated another: "Just because PFOA can cause kidney cancer doesn't mean that it caused Mrs. Bartlett's kidney cancer."

Then he said something that made me nearly jump out of my skin. "We threw the gauntlet down to Mrs. Bartlett's lawyers. Out of the eight million pages of records involved in this case, show me one document— just one—show me one document where anybody at DuPont said they expected that harm was likely to anyone in the community.

"And despite the many, many years of litigation you've heard about, more lawyers than I can count working against DuPont every day, digging through every nook and cranny, every file cabinet, every desk drawer, every computer of anybody who had anything to do with PFOA. . . .

"Ladies and gentlemen, you can search each and every one of those emails with a microscope and you will not find a single statement in any of them that they expected any harm to anyone."

I looked over at Pap, who was ready to launch out of his chair. We both knew about the secret weapon he had folded in his breast pocket.

I had pulled it some nights before, while rereading and digging back through a huge stack of our key documents (yet again). DuPont had been hammering the argument at trial that they had simply been following the advice of 3M—the manufacturer of PFOA for the first fifty years—and that they had never been told by 3M that there was any reason to suspect that there were any health concerns or risks from PFOA. So I had pulled my stack of old 3M documents and went back over all of them one more time. I came to one I had found years before and probably looked at a hundred times. Reading it again for the hundred and first time, I spotted something that had been in front of my eyes the entire time. Now it jumped out at me as if someone had shined a giant spotlight upon it.

"You know," I had told Pap the next morning, "I was looking through the documents again, and I found this." I'd handed him the document. "I think it might be useful."

Pap had looked it over, then glanced back at me and smiled as if he had just pulled the sword from the stone. We'd decided to flash it briefly in front of the last witness so it could be logged in the case as evidence but held off on emphasizing its import until our very last rebuttal. It would be the final surprise.

So now, when Damond Mace finished his closing and rested his case, Pap exploded back up before the jury as though his hair were on fire. He was flushed with emotion and adrenaline as he turned to the jury with a peculiar expression.

"Mr. Mace said to me: Show me a document."

Pap swiveled to glare at the defense table, where DuPont's counsel and their corporate scientist, Bobby Rickard, sat. He pulled the document from the pocket of his coat and unfolded it with great drama. "Mr. Mace, Mr. Rickard, there's the document."

It was a 3M Material Safety Data Sheet from 1997—providing a detailed description of the properties, exposure risks, and health hazards of a material 3M had sent to DuPont. He read from it: "Warning: Contains a chemical which can cause cancer." And the chemical that was being referred to? PFOA. So here was 3M telling DuPont about as explicitly as possible—in 1997—that PFOA could cause cancer.

Pap turned back to the jury, waving the document. We projected it onto the courtroom screen for everyone to see.

"Show me a document?" he said. "There you go. *Read THAT!*"

The DuPont lawyers looked at the document, unable to hide their dismay. I could see them reading the sheet intently, as if they thought that studying it hard enough might make those words, *cause cancer*, disappear.

Pap let his thunderclap resound briefly, then moved quickly through the denouement. "I hope DuPont hears me. I hope they hear that there's no system in this country that says that a corporation is too big to be held accountable."

I studied the jury, tried to read their faces. Was that a glimmer of acknowledgment? I was afraid to be too hopeful. You just never know what a jury will do.

Pap said, "You tell me that you're okay with the conduct of Mr. DuPont in the way he's treated these people all along the Ohio River. Because if it's okay, the system is broken. The cynic is right."

· · ·

At 4:00 p.m. on October 7, 2015, after less than a day of deliberations, we were told that the jury had a verdict. I was so nervous and anxious I thought I would be sick. This was it. Everything I had been working on—sixteen years of obsession—was all coming down to this one moment. We filed into the courtroom and took our seats at the counsel table with Mrs. Bartlett.

After what seemed like hours but was only a few minutes, we stood as the jury entered. After the jury was seated, the bailiff came in and raised the gavel. We stood again as Judge Sargus entered the courtroom. I worried that my legs wouldn't hold me. My nerves were on fire. I was terrified that the intense anxiety might trigger a seizure or that my overwhelming emotion would be otherwise visible to the jury.

"Good afternoon, ladies and gentlemen," Judge Sargus said to the jury, to us, and to a gallery that was even more packed than it had been on the first days of trial.

He asked the jury foreman to stand.

"I understand, sir, that verdicts have been rendered in this case?" he said.

"Yes, they have, Your Honor."

The very air sizzled. As instructed, the juror handed the verdict document to the deputy clerk, who stood and read it to the courtroom in a flat, matter-of-fact voice.

"Civil action C2-13-170, Carla Marie Bartlett versus E. I. DuPont. Jury verdict: Do you find in favor of Mrs. Bartlett on her negligence claim? Answer: yes; signed by the seven jurors."

I had to lower my head. I was afraid I was visibly trembling.

"If you found in favor of Mrs. Bartlett, what damages, if any, do you find Mrs. Bartlett is entitled to on her negligence claim? Answer . . ."

The deputy clerk paused and looked at the judge, who waved him over to the bench to confer privately. The deputy clerk showed the document to the judge, who nodded and allowed him to continue.

"Answer: $1.1 million; signed by the seven jurors."

Judge Sargus interrupted before he read further, turning to the jury with a question. "The writing is a little bit unclear. I'm going to ask all of you, is that the number correctly read?"

"Yes," the jurors said.

"Jury verdict form," the deputy clerk continued. "Do you find in favor of Mrs. Bartlett on her negligent infliction of emotional-distress claim? Answer: yes; signed by the seven jurors.

"If you found in favor of Mrs. Bartlett, what damages, if any, do you find Mrs. Bartlett is entitled to on her negligent infliction of emotional distress claim? Answer: $500,000, signed by the seven jurors."

A total of $1.6 million in compensation for the ways in which DuPont had injured her.

As the jury and judge exited the courtroom, I sat there, stunned and trembling. I knew that if I had to speak, I might cry. I stood to congratulate the team and particularly Mrs. Bartlett. Thankfully, words were unnecessary. She gave me a huge hug.

So many emotions. So many years. So much work. So many people. And in the blink of an eye, everything had changed. I would need some time to absorb all of it.

But time was not something we had yet. The next trial was scheduled to begin in just a few weeks.

34

THE RECKONING

December 18, 2015
Parkersburg, West Virginia

On a thirty-one-degree day in December, I stood in a field in West Virginia, shivering and miserable as I waited for a photographer to get his shot. Sarah had bought me a brand-new coat to wear for the shoot, but I had forgotten it at home. Out of habit, I had grabbed my old one, which was missing a button, covered with cat hair, and not all that warm. We had been out in the field for hours.

The photo shoot was for an upcoming media story about the PFOA saga. Like most interactions with the press, it made me nervous. Dealing with the press was usually a task I happily left to my co-counsels. When a journalist named Nathaniel Rich had called for a story he was working on for the *New York Times*, I had been reluctant to participate.

But when I thought about the fact that EPA still wasn't moving on PFOA regulation, that the rest of the country still didn't realize the likely contamination of their own drinking-water supplies, I reconsidered. A story explaining and highlighting the problem in the *New York Times* might bring greater awareness of the looming public health threat. It might help prompt EPA to finally do something. But I really had no idea what the *Times* had in mind. After much internal struggle, I eventually agreed to take the chance and work with Nathaniel. Speaking with him and doing the interviews wasn't bad, but standing in the cold in front of the camera was a special kind of agony.

With the permission of Jim and Della Tennant, we had come out to the home place, where the photographer wanted to take some shots of me in the location where the PFOA saga had begun. As I stood at the edge of an empty meadow where a herd of cattle had once grazed, it was hard not to think of Earl and Sandy. And how much had changed.

Behind me, Lydia Tennant's two-story house was vacant and dark. So was the barn and grain silo that the family had built by hand. In the early 2000s, Earl's youngest daughter, Amy, had been living in the house with her husband and two young kids. But after the problems with Dry Run Landfill, they had decided it was safer to move off the farm. Amy still suffered from terrible migraines, though, and the kids both had health problems.

· · ·

It was nice to be with Jim and Della again; they had accompanied me throughout the shoot and were now huddled in their minivan trying to get warm as the photographer set up his last shots.

Even after the majority of the Parkersburg community had rallied around the class action, even after the Bartlett verdict, there were still some die-hard DuPont loyalists who snubbed Jim and Della from time to time or left when they walked into a room. Tension remained between people who worried first about their health and people who worried first about the economy. You could still feel that tension in the grocery aisles of the Piggly Wiggly and in the glances between tables at the Western Sizzlin, where Jim and Della took me to eat when I was in town.

"I'm not trying to put economy over health, but if DuPont would close, people will leave," the mayor of Belpre had told a *Washington Post* reporter. "With this C8 case, no one wins, everyone loses."

Belpre was one of the six contaminated water districts in our class action, but Mayor Michael Lorentz said he still drank the water every day. He said he didn't know a single person with one of the six linked diseases and claimed that there wasn't "a neighborhood in Belpre that doesn't have a DuPont employee or retiree in it."

Mayor Lorentz still considered DuPont "a terrific neighbor" and "an asset to the community." But then, he had worked for chemical plants himself for most of his career, including twenty-six years at the local Shell plant.

"I believe in my heart that if I got sick, and they cut me open and found I was full of crude oil, that that corporation would take care of me," he said. "You get better results working with a company than by filing a lawsuit."

He had nothing against Mrs. Bartlett, he said, but he personally believed that anyone who was affected by PFOA "had underlying health issues."

"The C8 thing has been overdone," Lorentz said. And that wasn't all he knew for sure. He could also tell you who he thought benefited most from the litigation: "The lawyers."

Although a common misperception, that was far from my reality. The original class settlement had occurred in 2004, more than eleven years before. During those eleven years, we had been incurring enormous additional expenses through the cases in Parkersburg, New Jersey, and Minnesota and in connection with the extensive Science Panel work. With the new MDL phase that had started in 2013, I was spending even more time on the case than ever before and racking up even higher costs for the firm. Thousands upon thousands of hours were being billed to cases with rapidly escalating out-of-pocket costs and no corresponding income.

As a partner whose compensation depended to a large degree on my contributions to the profitability of the firm, costing the firm more money year after year after year had actually resulted in my take-home pay going down for several years in a row.

Yes, we had won the Bartlett case with a $1.6 million verdict, but the firm had still not been paid. Nor was it clear that we ever would be; DuPont was appealing the verdict. So I still stayed at the Red Roof Inn when I went to Parkersburg, still ate at the Western Sizzlin, and still drove my 1990 Toyota Celica with 200,000 miles on the odometer. I had to bite my tongue when I read the comments about all the lawyers "getting rich" off PFOA.

I wished that people could see the bigger picture: the decade of unbillable hours, the expense of reaching out to seventy thousand plaintiffs, and the costs of million-dollar trials. Even when there is a recovery, you have to repay all the unreimbursed expenses, divide what's left among the various firms, then divvy that up among all the partners of each firm. (I now had scores of partners at my firm, following mergers with firms in Indianapolis and Chicago.) So what's left is not nearly the pile of riches most people imagine.

I didn't get into all that with the *New York Times*. I was hoping to focus on the problem with PFOA—not the inner workings of big law firms. Nevertheless, the *New York Times* story seemed like a never-ending process, beginning with in-person interviews and ending with dozens of calls from fact checkers. And now it needed one photograph, which somehow took up a whole day.

By late afternoon, I was still standing in the field. My face, fingers, and toes were totally numb from the cold, and it had started to snow. I could not muster another smile.

"Just a few more . . ." *Click-click-click.*

I was exhausted, frozen, and done. The misery was written all over my face. Of course, *that* was the shot they chose to run.

I was floored to learn that the story not only would be running but would be the cover story in the *New York Times Magazine*. I was even more amazed when I saw the headline accompanying the story when it ran on January 6, 2016: "The Lawyer Who Became DuPont's Worst Nightmare."

But the most unexpected result of the story was the number of people who took the time to write and say thank you. My inbox was flooded with letters and emails—most of them from complete strangers. I made it a point to try to respond to and thank each person who had reached out to me. I was deeply moved. I printed the emails out as they came in so I could take them home and share them with Sarah and the boys. Soon I had a stack as big as a phone book. Sarah and I took them to a quiet neighborhood restaurant to read the messages over drinks and dinner. "If nothing else comes of this case," she said, "this alone made it all worthwhile."

The media exposure brought another revelatory response. I was floored by the number of partners at my own firm who approached me with a sense of surprise and curiosity. Many of them had joined the firm only recently through one of the mergers and had not fully understood the scope of what I'd been working on all those years.

My longtime boss, Tom Terp, was now the managing partner of Taft. He framed a copy of the cover and the story and presented it to me in a partner meeting.

I liked it even better than the Teflon frying pan.

. . .

With the Bartlett verdict, DuPont had, for the first time, been held directly legally liable for the PFOA mess. That bellwether outcome—including the specific damages assigned by the jury—provided the information DuPont had been saying they needed to assess the rest of the personal injury cases.

Compared to other members of the class, Carla Bartlett had a relatively low PFOA blood level. She lived in one of the communities farthest away from the Washington Works plant, with some of the lowest PFOA drinking-water levels. Her cancer had been caught early (in stage 1). After surgery, she had had no further complications, had not required chemotherapy, and had been cancer free for more than twenty years. Because the hospital had taken her on as a charity case, she had had no out-of-pocket medical expenses.

If DuPont had lost this case—a less dramatic one than some others—and a jury had awarded $1.6 million in compensatory damages, that should have sent a strong message about the strength of the underlying facts supporting the company's liability. Yet, less than twenty-four hours after the jury verdict, both DuPont and Chemours told the press and the public that the outcome of the case had actually *validated* their position. DuPont's spokesman, Gregg Schmidt, spun it this way: The lack of an additional punitive damage award from the jury "validates DuPont's position that at no time was there a conscious disregard for those living near DuPont's Parkersburg, West Virginia, plant. DuPont believes it has always acted reasonably and responsibly in its handling [of PFOA], acting on the health and environmental information that was available to industry and regulators at the time the company used [PFOA]."

Chemours' representative, Janet Smith, echoed DuPont: "We maintain that DuPont acted reasonably and responsibly at each stage in the long history of PFOA, placing a high priority on the safety of workers and community members."

According to both companies, the jury's finding of any responsibility or damage award was simply in error and would be appealed.

The full story had finally been shown to a jury. Its members had seen what Earl had tried so desperately to bring to light nearly two decades earlier. They had understood all the things I had been shouting into the wind for years. And they had gotten the point.

It seemed that now the only one that still refused to see or accept the truth was DuPont. If it was going to take punitive damages to convince them otherwise, we'd double down to get those in our next trial.

. . .

In the summer of 2016, we took our next bellwether case before a jury. David Freeman was a college professor who had survived testicular cancer. The operation that had saved his life had required the removal of one testicle as well as several lymph nodes in his abdomen, requiring a surgery at least as severe as Carla Bartlett's.

We had the advantage of essentially presenting the same foundational facts about PFOA and DuPont's decisions. So we were able to fine-tune our strategy, polish our presentation, and adjust the framework around Mr. Freeman's particular circumstances. This time, though, we added emphasis to the facts showing DuPont's conscious disregard of the risks in the hope of adding punitive damages to any award.

On July 5, 2016, I was once again parked behind the plaintiffs' table, watching Pap deliver his portion of our team's closing statements. Though

Mrs. Bartlett's case had overwhelmed me with emotion, this one struck me with the gravity of what we had done. Building this case had been like building a massive engine with a thousand moving parts. Everything had to work together perfectly to hold DuPont accountable. One broken part might have jammed the whole thing. One missing piece could have ruined us.

Pap was one of those parts, and today he wasn't just moving, he was rolling.

"Yesterday," he began, making eye contact with the seven jury members one by one, "America celebrated its independence. And part of that celebration should center around the idea of what we gained from that independence. What came out of that was the idea that there is no company that's too big—there's no king, there's no president, there's no entity that's too big for an American citizen to bring into a courtroom and say: 'What you have done is wrong.'"

He then introduced the villain of his saga. "I call him the puppet master of the Ohio River Valley—that's what Mr. DuPont is."

In a court of law, a corporation is treated like a person. It has the same legal rights under the Fourteenth Amendment as a human being: equal protection under the law, the right to due process in a court of law. Also like a person, you might argue, a corporation has its own values, volition, and conscience.

"In order to come into the courtroom and even tell the story that they told, as unusual as it is, they had to be able to manipulate the community over the years, environmentalists over the years, their own employees."

I thought about all the spinning of information and facts on so many fronts for so many years. It had been nearly impossible to spot in a system so complex, so dynamic. And when I did spot it, did begin to ferret out the truth, I felt they had attempted to block me through a questionable collaboration with the state. They tried to hold back their most damaging documents in discovery for as long as possible, obscured the issue through confusing water-testing measures, questioned the Science Panel they had agreed not to challenge, contested the general causation issue they were contractually obligated not to contest.

"They had to manipulate both sides of the river. They had to manipulate scientists. They had to manipulate doctors who needed this information to treat patients but didn't have it. They had to delay. They had to cover up. In order to cover up, ladies and gentlemen . . . you have to be equipped to do that. You have to have some very good lawyers. You have to have some very good PR people to carry on the cover-up. You have to have, more importantly, a strategy."

We were going to make sure that punitive damages were awarded this time. We were going to make DuPont's conscious disregard of the risks crystal clear for this jury. This time, in addition to showing them just how much DuPont's talented scientists had known about PFOA—and how the business had ignored them—we would hammer hard on the company's efforts to obscure and misdirect the public and the regulators. With the documents I had discovered, there should be no question about their intentions.

Remember the "connect the dots" document, Pap told the jury; use that as "your North Star. This is their words, not mine: They say, *Is there a strategy we can use to minimize the amount of information being disseminated?* What is it they [don't] want to disseminate? They don't want people to understand that they are poisoning the river with fifty thousand pounds of toxin a year that they know has the capacity to cause cancer. They don't want to disseminate the fact that the PFOA that is a poison stays in people's blood for twenty to twenty-five years."

Hearing that summation, I had an epiphany. Every painful moment of the past seventeen years—including the moments that had felt like defeat—had all been necessary to get to this point. Even the frustrating setbacks, such as the recent lawsuits in Minnesota and Parkersburg, both of which had failed to be certified, and the relatively small $8.3 million class settlement in the New Jersey case, were absolutely essential. Even though their outcomes had been disappointing, those lawsuits had allowed me to continue getting documents from DuPont and numerous other sources through active discovery for many additional years. And many of those documents were instrumental in the trials now taking place.

Including "connect the dots." It was the untold thousandth document that had come to me, but for the purposes of this trial, it was document number 283, and it was the key.

"If you want to understand the technique that this company used to cover everything up, you begin with document 283, and it's going to show you," Pap said. "Anything that limits the interest in dots will tend to diminish our exposure. They understand that if the public finds out, there's going to be discontent. They understand if there's discontent, it's going to increase their liability. It's going to increase their *exposure*."

DuPont had hoped that nobody would see the dots, or, if they did, that they wouldn't be able to connect them. But Earl had seen the dots. Earl had connected them. I wasn't DuPont's worst nightmare. Earl was. Earl had taught me what courage means. Even when the government, the community, even some of his own friends and neighbors had been pressuring him to back down, he had been the one to stand alone and throw stones at that corporate giant. He had pushed me to keep fighting even when his battle was done. Because he had known this wasn't just about him.

Right now, Pap was making sure that the jury knew the matter at hand was all about DuPont. "They say, what is the number of potential plaintiffs in this case?" He waved a document that was projected on the screen. In it, DuPont referred to those who had been drinking their contaminated water as "receptors."

"Understand," Pap intoned in his magnetic drawl, "every [time] you see 'receptors,' they're talking about people. They're talking about mamas and daddies and children. Those are 'receptors.' They're not interested in keeping people healthy. They're trying to figure out how much this is going to cost them after fifty-seven years of doing it. The point is, all this was avoidable. It was avoidable for many, many people in the Ohio Valley. All they had to do was just do the right thing."

The company had known how to destroy the molecule. They had known that once it got out into the environment or into people's blood, it could not be destroyed. But it could destroy. The genie was out of the bottle. The toxin was now in the blood of every person on Earth—in living things in every corner of the planet. They had contaminated the entire globe.

"So as I connect the dots, I'm going to use this." Pap paused, letting suspense build like an electric charge in the silent courtroom. Then he spat out the word, *zap!* "Malice," he said. "Malice is a word you're going to hear. Malice, as you're hearing the word, you might say, I think malice is these things. I think it's hate and it's spite and it's ill will and it's recklessness and premeditation and evil. It's none of those things. Malice, in the terms of the way that we use it in this case, and the way that you'll be asked to evaluate their conduct, is not 'was it evil,' not dislike or hate or any of those things. It is simply that actual malice is *conscious disregard.* You're conscious of something and you disregard it and it ends up hurting somebody. That's what it means. That's the law."

I watched the jury members take in Pap's bravura performance, and I hoped they understood what we had asked them to understand.

This was no accident. This was a choice.

On July 6, 2016, the jury found in favor of David Freeman, awarding him $5.1 million in compensatory damages.

This time, though, it added $500,000 in punitive damages.

This time it found actual malice.

That, I thought, should finally send DuPont the message they needed to hear, the sign that it was time to fix the problem once and for all.

DuPont and Chemours promptly announced that the case would be appealed.

I could smell tobacco in the air.

. . .

So we pushed on to the third MDL trial. That case, for Kenneth Vigneron, continued the trend of escalating damages. Mr. Vigneron, a fifty-six-year-old truck driver and a father of four, had an even more traumatic case of testicular cancer.

The jury found in his favor, awarding $2 million in compensatory damages and *$10.5* million in punitives.

Surely, this third case would be the charm.

Nope.

Again we heard from DuPont and Chemours that the court and jury had simply erred and all would be fixed on appeal.

The bellwether and trial process seemed to be getting us nowhere. Trial after trial was ending the same way: DuPont was held liable, with ever-increasing damage and punitive awards, yet they vowed to fight on. At this rate, with more than three thousand cases to go, we could be at it for hundreds of years. Most of our clients could be dead before their cases were ever heard.

Judge Sargus decided to speed up the process. If DuPont was intent on trying all those cases, we would need to start setting many more trials much more quickly. Soon an order was entered in the MDL scheduling forty more cancer trials to be completed by the end of 2017, to be divided up among various judges and federal courts in Ohio, Kentucky, and West Virginia.

In the middle of the first of those trials, DuPont folded. Their lawyers agreed to resolve all the pending cases once and for all.

In February 2017, DuPont agreed to pay $670.7 million to settle the more than 3,500 lawsuits pending in Ohio and West Virginia, including the cases that had already gone to trial. Even though DuPont had fully briefed and argued all their appellate issues and were simply awaiting the appellate court's rulings, they voluntarily dismissed their appeals and settled those cases as well. Determining who would get what portion of the settlement amount would be a complex process, left to a special master and claims administrator appointed by the judge and overseen by the court. Important distinctions within the group, such as which disease or diseases were at issue, what medical procedures (such as chemotherapy) had been involved, how many years of agony and debilitation had been suffered, and whose life savings had been wiped out by massive medical bills would be assessed and ranked, and individual award amounts would be determined for each plaintiff.

I will never feel completely satisfied that neither Karen Robinson's son, Chip, nor Bucky Bailey were parties to the MDL settlement, as birth defects were not included by the Science Panel as being probably linked to PFOA. I find it hard to believe that future research won't confirm that there is in fact such a connection. But for the more than 3,500 individuals who did participate, I feel a great sense of vindication. These included Sue Bailey, who suffered from a linked thyroid disease, and Ken Wamsley, who had linked ulcerative colitis.

With the case finally over, Ken sat down to try to sum up all the ways his life had been impacted. "I can give you a list of my health problems," he wrote. "Ulcerative colitis. Aggressive rectal carcinoma. Surgery to remove my spleen. Teeth that turned black and rotted out. But I'd rather show you what all those things mean. I want to show you everything I gave to DuPont, and everything they have taken from me.

"Every morning I wake up and remember: I don't have a rectum. The place where it should be is all sealed up. My excrement dribbles out of a hole in my stomach. It pours into a pouch that's stuck to my belly, right next to two large scars where they cut me open twice. . . . I know other people live with colostomy bags, but it has stolen my independence. My confidence. My dignity. I've always been a go-getter, a people person, a leader. Now I'm so vulnerable and afraid that I rarely leave the house.

"I always dreamed that in retirement I'd be teaching math at the local community college. Or coaching a baseball team. Or traveling, seeing the world. Or finding someone to love and share a life with. Those dreams have been stolen from me—along with so many of life's simple pleasures: chewing a steak. Sipping a beer. Sitting on the toilet and reading the paper. I haven't sat on a toilet in fifteen years.

"I imagine other people have lost similar things. But here's why my suffering is different: it was inflicted by the company I devoted my life to. The company I trusted. The company that exposed me to *known* dangers and promised me I was safe. The company that lied to me, destroyed my health, and took away my faith in human decency. The company that continues to kill me, slowly but surely, from the inside out. The company I stood up for. The company I loved."

Knowing that our years of toil had brought some measure of compensation, however inadequate, to people who had suffered so intensely was humbling. These people had counted on me, and though it had taken a very long time, in the end, I hadn't failed them.

. . .

One day in 2016, my office phone rang with an unfamiliar number.

At first I strained to understand the accent of the woman on the other end of the line. She was a journalist calling from Dordrecht, in the Netherlands, home to a former DuPont (now Chemours) plant, the European sister of Washington Works. Dordrecht was one of the oldest and prettiest port cities in the Netherlands, a maze of canals and cobblestoned streets nicknamed "Little Amsterdam." But it had an ugly problem.

The journalist was investigating the chemical replacement for PFOA. Introduced quietly by DuPont in 2009 during EPA's ten-year PFOA phase-out program, it was an alternative called GenX. Its structure was almost identical to that of PFOA, only with two fewer carbon atoms in the backbone.

GenX appeared to have gotten into the Dordrecht water. But no one knew what that meant. What was this stuff? Was it dangerous? Few people seemed to have even heard of the chemical at all. Chemours was saying there was no evidence of any adverse human health effects, but people were worried. They wanted it out of their water and wanted to know what it could do to them after drinking it all these years, but no one knew quite what to do.

"Mr. Bilott," she said. "Can you help us?"

. . .

That was just the start. After the *New York Times Magazine* piece and the news of the $670.7 million settlement, hardly a day passed without new inquiries about legal representation or new invitations to speak. I gave talks at universities and law schools, gave TV interviews, testified before the New York State senate, and delivered speeches to standing-room-only crowds in Italy, where the community organized a ten-thousand-person march demanding strict regulation of PFOA and related compounds grouped under the PFAS heading. I was invited to appear at the Sundance Film Festival premiere of a feature documentary, *The Devil*

We Know, about our West Virginia litigation, and had a chance to answer questions after the showing. I even traveled to Stockholm, Sweden, to receive the Right Livelihood Award, also known as the "Alternative Nobel Prize," for my work on PFOA, and met with assorted European leaders and representatives of the United Nations to discuss the spreading global chemical threat of PFOA and related chemicals.

I was painfully aware that just as I could have finally been spending more time with Sarah and the boys, I became as unavailable as I had been when I was buried in trial preparation and spending weeks in a hotel in Columbus, Ohio, for the cancer trials. But Sarah encouraged me to do what I needed to do. She'd long been accustomed to our living parallel lives for long stretches, and she was entering an exciting new phase of her own. With the boys in high school and me always at work, Sarah had revived her legal career with a job as legal counsel for a local nonprofit called ProKids, dedicated to aiding children in crisis. And the boys, eyeball deep in the social and extracurricular whirl of active teenagers, would barely have noticed me if I had been around.

I didn't love spending most of 2017 in airports and hotel rooms, but I felt great satisfaction that I was getting the word out, not only nationally but internationally, about the still unresolved threat to millions of people who only wanted to see clean and safe water flowing from their taps. There was a growing awareness that the entire class of PFAS chemicals— as many as four thousand related compounds—might be a problem, including the newer replacement chemicals (like GenX) that were being billed as less persistent than PFOA and PFOS. Their structural similarity and some evidence from animal studies suggested they could be toxic or carcinogenic, but we were hearing the same familiar argument that nobody had done the extensive science required to reach a firm and actionable conclusion on the human impact of these additional PFAS chemicals.

As public awareness spread, so did testing of public waters for contamination. With increased testing, an ever-expanding map of PFOA/ PFAS contamination sites was revealed. Meanwhile, the Department of Defense began alerting communities that the drinking water near hundreds of military bases and airports was likely contaminated by PFOA released by the use of Class B firefighting foam to extinguish petroleum-based fires. Investigations are taking place around the world at nearly one thousand different sites with likely foam impacts.

A survey by the Environmental Working Group documented PFAS pollution at more than seven hundred sites in forty-nine states, including military bases, airports, manufacturing plants and landfills, and fire training sites. PFAS-contaminated tap water was now known to be impacting more than 100 million Americans—just over a third of the entire population. And studies around the world have continued to raise the possibility of links to health concerns far beyond the six probable links to diseases found by our C8 Science Panel; the most eye-catching of which for the media was a study of young men in Italy concluding that those who had been exposed to high levels of PFOA and PFOS had reproductive deficits, including "shorter penises, lower sperm counts, [and] lower sperm mobility."

Even with widening public attention on PFAS, and despite increasing political and regulatory pressure, it became painfully clear to me that nothing was actually being done to fix the problem. On May 20, 2016, just four months after Nathaniel's *New York Times Magazine* story on me—but almost exactly ten years after EPA's PFOA phaseout deal with DuPont (the deal in which DuPont had asked for a pause on any federal regulation)—EPA announced the first long-term federal health advisory level for PFOA in drinking water. It now claimed that 70 parts per trillion (0.07 ppb) was safe for lifetime exposure to both PFOA and PFOS and announced big promises to take action on PFAS. More than a year later, nothing had come of those promises.

In the spring and summer of 2018, EPA, under further intensifying public, media, and political pressure, announced that it was developing a "national plan" to manage PFAS pollution, which the agency said would involve examining "everything we know about PFOA and PFOS in drinking water." More than seventeen years after my letter warning of the dangers of PFOA, the agency claimed to finally be "beginning the necessary steps to propose designating PFOA and PFOS as 'hazardous substances'" and to be "currently developing groundwater cleanup recommendations for PFOA and PFOS at contaminated sites." It also promised that it was "taking action in close collaboration with our federal and state partners" to develop toxicity values for all the new PFAS chemicals with as-yet unknown risks.

And yet there was no further progress from EPA by the end of 2018. Then, in February 2019, EPA announced with great fanfare that it had "begun the process" of finally listing PFOA and PFOS as hazardous substances under the Superfund law. That process has been "beginning" for almost two decades. Delaware senator Tom Carper echoed my thoughts exactly when he said, "It has taken the EPA nearly a year just to kick the can even farther down the road."

To get meaningful action for the millions of Americans affected by PFAS chemical contamination, or simply to confirm the extent of the health threat—including cancer—I realized that it was necessary to resort once again to the judicial system: good ol' common-law tort concepts. I focused on a single, terrifying fact: these chemicals are in everybody's blood. The victims aren't isolated in any single community, region, or state. They are all of us.

The claim would be simple: companies had wrongly put their chemicals into the environment, knowing we would serve as human sponges to soak up and store their toxins in our bloodstreams like millions of free, unpermitted, walking landfills; and that ubiquitous blood contamination—a ticking time bomb, placed in our bodies without our consent or knowledge—was, in itself, the injury we would be suing over. The remedy sought would not be cash. It would be more basic than that: scientific knowledge.

My idea was to bring a new lawsuit against the companies who created, used, and profited from these chemicals to fund whatever science was necessary to confirm the human health risks from the broader class of PFAS chemicals, the science that should have been done before any of these chemicals were ever put into commercial production. We would be seeking to create an expanded version of our C8 Science Panel of unbiased scientists approved by both sides to determine, once and for all, what this blood contamination means for us with regard to this larger class of PFAS chemicals. Our suit would seek to force the companies, through settlement or judgment, to pay for this research, and to be required to accept that once a determination was made, it would be the final word. Inarguable and un-appealable. After all, why should the taxpayers be paying to understand what these companies did to them?

Even after twenty years as a plaintiffs' attorney, some of the biggest recoveries in my firm's history, and an "Alternative Nobel Prize," I was no less nervous about presenting this idea to my firm than I had been with the earlier cases. If an individual property damage and personal injury case could take three years, and pursuing claims on behalf of a small community stretched out for sixteen more, then how long—and how many unbilled hours—would a class action involving 326 million Americans require? Once again, I found myself asking my partners to jump into the by now well-worn rabbit hole with me. After lengthy and exhaustive study, they agreed.

On October 4, 2018, I filed a class action claim against eight chemical companies on behalf of the PFAS-contaminated American People in the United States District Court for the Southern District of Ohio, Judge Sargus presiding.

I have a lot of work to do.

EPILOGUE

Frustrated by the lack of action or any firm deadlines to do anything about PFAS, senators and representatives have introduced bipartisan bills in both the US House and the Senate that would require federal government regulation of PFAS in drinking water by a specific date, though the legislation has not yet passed. Other bills were introduced in early 2019 to require EPA to designate PFOA and PFAS "hazardous substances" under the federal Superfund statute by a specific date, but that legislation also has not yet passed. As for the dozens of other PFAS chemicals, EPA promises to "close the gap on science as quickly as possible."

Acknowledgments

My deepest, warmest, and most heartfelt thanks, gratitude, and love to my phenomenal wife, Sarah, and our amazing sons, Teddy, Charlie, and Tony, who lived through this decades-long saga and supported me throughout the original events as they transpired, and then through the reliving of it all as this book was created, along with my parents, Ray and Emily; sister, Beth; brother-in law, Terry; and all my and Sarah's extended family, including my nephews and niece and Sarah's parents, siblings, nephews, nieces, aunts, uncles, and cousins. A special thanks to Grammer, who first connected me to the people and events described herein that would ultimately change my life in so many ways.

My humblest thanks and appreciation to the tens of thousands of people in West Virginia, Ohio, Minnesota, New Jersey, and other communities across the country who bestowed upon me the honor of representing them on PFAS issues over the last two decades; not to only all the individuals who graciously shared their personal stories for this book, including the incomparable extended Tennant family (including Earl, Sandy, Jim, Della, Jack, Amy, Crystal,* Martha, and Terry), the Kigers (Joe and Darlene), the Robinsons (Karen and Chip), the Baileys (Sue and Bucky), Ken Wamsley, Carla Bartlett, David Freeman, and Ken Vigneron, but to all the individuals and their families who supported and participated in the massive C8 Health Project by providing personal health

---

* As the proud daughter of a farmer, Crystal Tennant would like people to be aware of the Future Farmers of America organization. You can learn more by going to www.ffa.org or www.blennerhassettffa.theaet.com.

373

information and blood samples, all of which became critical data that helped the entire world finally see the hidden risks posed by PFOA.

My sincerest thanks to everyone at my law firm, Taft Stettinius & Hollister, LLP, both past and present, who provided me the opportunity and support to wade into uncharted waters and take on—and keep pursuing—issues of global contamination, along with all the attorneys, paralegals, clerks, and others at all the law firms that were vital in helping us not only pursue this work but succeed at it, including not only those specifically mentioned in the book (such as Kathleen Welch, Tom Terp, Kim Burke, Mike Papantonio, Mike London, and Gary Douglas) but also those not specifically identified (such as Gerry Rapien, Steve Justice, David Butler, Kevin Madonna, Rebecca Newman, John Nalbandian, Aaron Herzig, Robert F. Kennedy Jr., Deborah McCrea, Rob Craig, Sue Koester, Tim O'Brien, Jeff Gaddy, Wes Bowden, Carol Moore, and Ashley Brittain Landers). A special thanks to Ned McWilliams for helping to elevate awareness of the PFOA story to new levels, and particularly to Larry Winter, my original co-counsel in the litigation that started all of this, whose guidance, loyalty, and unwavering support helped me to keep going through many difficult years.

I'm not sure I, or anyone, could ever properly thank the amazing team of talented, innovative, and dedicated professionals, including but not limited to Dr. Paul Brooks, Art Maher, Troy Young, and Patsy Flensborg, who came together and designed, implemented, and successfully completed the unprecedented C8 Health Project, which became one of the biggest and most successful human health studies ever undertaken; nor are there words adequate to express my appreciation for the work of the C8 Science and Medical Panelists and all the consultants and scientists who helped them interpret, analyze, confirm, and monitor the human health risks from PFOA.

I am beyond indebted to and grateful for the advice, education, and assistance provided by the various medical doctors, epidemiologists, risk assessors, analytical chemists, toxicologists, and other scientists, such as Drs. David Gray, Richard Clapp, Barry Levy, James Dahlgren, and James Smith, who helped me early on to understand the complex and often confusing and dizzying world of PFAS chemicals, along with all the scientists and researchers who have used their training and talents to speak out and help educate and warn the scientific and regulatory communities and the public about the dangers of PFOA and PFAS chemicals, including but not limited to Drs. Fardin Oliaei, Glenn Evers, Richard Purdy, and Arlene Blum.

A very special thanks for the tireless, dedicated efforts of everyone at the Environmental Working Group—particularly Ken Cook—who have done remarkable, unsurpassed work to elevate awareness of the risk and threats of PFOA and PFAS. EWG's work on this issue has been truly phenomenal. We all owe them a debt of gratitude.

I also want to thank all the other NGOs, citizens' groups and organizations, journalists, writers, and media members who have helped alert the public to the global health threat presented, including but not limited to the Green Science Policy Institute, Ken Ward, Sharon Lerner, Callie Lyons, Mariah Blake, and Nathaniel Rich.

A special thanks also to Bill Couzens and my fellow board members at Less Cancer and the folks at the Right Livelihood Award Foundation for continuing the effort to spread awareness of the cancer threat posed by PFOA and other PFAS chemicals.

My deepest appreciation is also extended to the entire extraordinary team at Atria Books/Simon & Schuster for caring so deeply about my story and this issue. I especially want to thank my wonderful editor, Peter Borland, whose deft and insightful editing and unwavering belief in this book made it possible. A huge thanks also to his assistant, Sean deLone, who along with Peter worked tirelessly and no doubt went above and beyond, and additional thanks to Libby McGuire, Benjamin Holmes, and James Iacobelli. An additional thanks to my wife, Sarah, for her thoughtful review, edits, and insights.

My most sincere appreciation and thanks are also extended to Tom Shroder, whose incredible literary skill and talent were instrumental in weaving together the often esoteric, complex, and confusing parallel worlds of science, regulation, law, politics, and corporate culture. Thank you, Tom, for your patience, your compassion, and for bringing this story to life within the pages of this book. I will also be forever grateful for the early contributions and Herculean research of the truly amazing and talented Kim Cross. Thank you, Kim.

Ultimately, this book would never have happened without the unmatched support, passion, and tireless effort and work of my literary agent, Laurie Bernstein, from Side by Side Literary Productions, Inc. Her phenomenal perseverance, dogged tenacity, encouragement, talent, and unwavering belief in the power and importance of this story made this book a reality. Thank you, Laurie.

Index